Abstraction and Infinity

Abstraction and Infinity

Paolo Mancosu

Great Clarendon Street, Oxford, OX2 6DP,
United Kingdom

Oxford University Press is a department of the University of Oxford.
It furthers the University's objective of excellence in research, scholarship,
and education by publishing worldwide. Oxford is a registered trade mark of
Oxford University Press in the UK and in certain other countries

First Edition published in 2016

Impression: 2

Published in the United States of America by Oxford University Press
198 Madison Avenue, New York, NY 10016, United States of America

British Library Cataloguing in Publication Data

Data available

Library of Congress Control Number: 2016945125

ISBN 978-0-19-874682-9

Printed in Great Britain by
CPI Group (UK) Ltd, Croydon CR0 4YY

To my parents, Angela and Porfidio,
whose love and support have made
so many things possible

Contents

Introduction

As the title indicates, the book is an investigation into abstraction, infinity and their interaction.

Abstraction is of course "said in many ways" but the meaning relevant to my investigation is that which has had pride of place in the last thirty years in philosophy of mathematics, namely the meaning of abstraction that is connected to the use of abstraction principles in neo-logicism.

Infinity is, appropriately, a never-ending concern for the philosopher of mathematics. Since the Cantorian revolution we have witnessed a number of fascinating developments, first in the study of non-Archimedean systems and, most recently, in the related development of theories of numerosities for infinite sets that satisfy the part-whole principle.

While the topics of abstraction and infinity, as briefly defined so far, might appear to have almost no intersection, we will see that the connections are profound. The most familiar abstraction principle is Hume's Principle. Hume's Principle assigns numbers to concepts according to the criterion that two concepts have the same number if and only if the objects falling under each one of them can be put in one to one correspondence. The principle assigns numbers to infinite concepts (i.e. concepts with an infinite extension) using the same criterion that Cantor used in his theory of cardinal numbers. Recent developments in mathematics have witnessed the emergence of a theory that preserves a different assignment of "size" to infinite concepts, namely one that satisfies the part-whole principle (if a set A is strictly included in a set B then the numerosity of A is strictly less than the numerosity of B). The latter theory, known as the theory of numerosities, opens up the conceptual freedom to question whether Hume's Principle is the only or the correct option for the neo-logicist and consequently also invites a different perspective on the discussion concerning the status of Hume's Principle as analytic or as a conceptual truth.

I will now explain in more detail the background to the work on abstraction, infinity and their interaction presented in this book.

Abstraction

One of the most influential programs in contemporary philosophy of mathematics is the neo-logicist program.[1] At the core of neo-logicism are a technical result and a set of philosophical considerations. The technical result is called Frege's theorem: second-order Peano arithmetic can be derived from Hume's Principle and second-order logic. A cluster of philosophical considerations are aimed at showing that Hume's Principle is analytic and that second-order logic does not take us out of the realm of analyticity. As is well known, there are abstraction principles that are inconsistent (the notorious Basic Law V in Frege's *Grundgesetze*) and abstraction principles, such as the Parity Principle or the Nuisance Principle, that are true on a domain if and only if Hume's Principle is false on that domain. The attempt at differentiating 'good' from 'bad' abstractions has given rise, in the last twenty years, to an extensive literature on abstraction principles.[2] This literature is mathematically sophisticated and philosophically stimulating.

It is however surprising that apart from some perfunctory references to Frege's *Grundlagen der Arithmetik*, §64, the analytic literature on abstraction principles ignores the extensive discussion on such principles that occupied many mathematicians and philosophers at the end of the nineteenth century and at the beginning of the twentieth century, including Peano, Burali-Forti, Russell, Padoa, and Scholz. The discussion was framed within the context of so-called "definitions by abstraction" (the term appeared in print in 1894 used by members of the Peano school but the idea occurred earlier). An example of such a definition is given by Frege in §64 of the *Grundlagen*: for all lines *a* and *b*, the direction of line *a* is equal to the direction of line *b* if and only if *a* and *b* are parallel. But Frege's wording in the *Grundlagen* seems to imply that the practice of defining by abstraction was still considered quite unusual when Frege was writing the *Grundlagen*.

In the first two chapters of the book I show that the logical discussion of definitions by abstraction is rooted in a stable and widespread mathematical practice and that reflection on the mathematical practice led to important foundational issues.

The first chapter is devoted to showing that, contrary to what Frege intimates in the *Grundlagen*, abstraction principles were quite widespread in the mathematical practice that preceded Frege's discussion of them. I cite extensively from nineteenth century sources in number theory, geometry, algebra, vector theory, set

---

[1] See Hale and Wright (2001*a*) for a classic source and Tennant (2013) for an overview.

[2] See Linnebo (2009*a*) for a recent issue of *Synthese* devoted to this problem. See also appendix 1 to Chapter 4 for a brief introduction to 'bad company' objections and criteria for admissible abstractions.

theory and foundations of the number systems, in order to show the widespread use of abstraction principles in the mathematical practice of the time and to articulate the foundational problems such principles gave rise to.

The second chapter, which builds on the first, is divided into two parts, the first of which analyzes Frege's discussion of abstraction principles in section 64 of the *Grundlagen*. In my investigations on Frege, I have been able to provide for the first time a contextual analysis of this pivotal section of the *Grundlagen* by rooting Frege's concerns in the geometrical tradition originating with Hermann Grassmann and in the tradition of attempts aimed at proving the parallel postulate by making the notion of 'direction' prior to the notion of parallelicity. The second part of Chapter 2 gives an extended analysis of the Peano school (Peano, Burali-Forti, and Padoa, among others) and Russell and shows how in their considerations on abstraction principles they foreshadowed many of the philosophical and foundational concerns that have been at the center of attention in the neo-logicist discussion.

Taken together these two chapters provide the first historical account of the mathematical practice of abstraction principles, their relevance for a better understanding of Frege, and the philosophical and foundational problems such principles posed in the nineteenth and early twentieth centuries.

Although infinity does appear in some of the preceding developments, for instance in the definition by abstraction of cardinals and ordinals, it is in Chapter 4 that I will specifically focus on the interaction between abstraction principles and infinity. The source of that interaction is however to be found in Chapter 3, to which I now turn.

Infinity

While during Greek times the infinite appears in mathematics mostly in heuristic roles, with the seventeenth century it begins to assert its right of existence alongside the finite. Desargues's points at infinity in projective geometry, Cavalieri's infinitary constructions and indivisibilist theories, Torricelli's infinitely long solid, Wallis's arithmetic of the infinite, Leibnizian differentials and Newtonian infinitesimal increments (moments) of time: these are just some of the major new approaches to infinity that characterize the seventeenth century. The shaky mastery of infinity gave rise to serious worries occasioned by the paradoxes, in the sense of outright contradictions and/or results that defied belief: Galileo's paradox about the comparison between the natural numbers and their squares; paradoxes in the theory of indivisibles; finite solids equal in volume to infinitely long ones;

quantities different from zero but smaller than any other quantity; confusions about the sums of infinite series. And so on.[3] The eighteenth century and the early nineteenth century reacted by attempting to eliminate infinity from mathematics as much as possible. In particular, the calculus became the major area of focus in the attempt to eliminate the problematic concepts originating from the fruitful but treacherous recourse to infinitary objects and constructions.

In the early nineteenth century, only a few dissenting voices seemed to embrace infinity in mathematics, the most notable one being that of Bolzano, whose *Paradoxes of the Infinite* (1851) had been prepared by years of engagement with the infinite in mathematics, physics, and philosophy. But the infinite in mathematics, which had been to a certain extent eliminated by rigorous theories of limits in the infinitesimal calculus, resurfaced in bolder and more challenging forms. By the middle of the nineteenth century, Dedekind and Riemann were already working with highly infinitistic constructions that ushered the development of set theory as a mathematical discipline in the hands of Dedekind and Cantor.

The development of Cantorian set theory marks a new dramatic phase in the history of the concept of infinity—a phase whose only comparable antecedent was the infinitistic revolution of the seventeenth century. Cantor's set theory, and Dedekind's set-theoretic constructions, pervaded many areas of mathematics (algebra, analysis, number theory, topology, etc.). As a consequence, set theory eventually came to play the role of a universal language of mathematics (a role that in the last half century has been only slightly eroded by category theory). This is not the place to recount the story of the emergence and eventual domination of infinitary set-theory (see Ferreirós 2007 for an excellent presentation) with all the logical, foundational, and philosophical issues that accompanied what is certainly one of the most impressive intellectual odysseys experienced by human thought. Rather, I would like to stress that a third infinitary uprising (a substantial event but not comparable to the sea change provoked by the infinitary revolution of the seventeenth century and that associated with the emergence and dominance of set theory) is related to non-Archimedean mathematics.

Archimedes' axiom states that given any two homogeneous quantities, say A and B, it is always possible to add A to itself a finite number of times (or B to itself a finite number of times), so that the new quantity is greater than B (or greater than A, respectively). Of course, many of the quantities considered in seventeenth century calculus, say Leibnizian infinitesimals, did not obey this Archimedean property. An infinitesimal can be added to itself any finite number of times but the outcome of that process will never be greater than any finite

---

[3] For a detailed exposition of the above topics see Mancosu (1996), especially Chapters 2, 5, and 6.

quantity, however small. With the elimination of infinitely small quantities from the theory of the infinitesimal calculus due to Bolzano, Cauchy, Weierstrass and others, non-Archimedean quantities in the calculus were relegated to engineering practice for a long time. In the 1960s, Abraham Robinson presented his theory of non-standard analysis that received a great deal of attention from philosophers and mathematicians.[4] Robinson's theory had given *droit de cité* to infinitely small and infinitely large quantities in the reconstruction of the infinitesimal calculus, now developed according to rigorous model-theoretic techniques. For a long time, Robinson's work was hailed as the first successful effort to develop a system of non-Archimedean quantities. Philip Ehrlich, in a series of fundamental papers, has argued that this widespread perception is in need of serious questioning.[5] Indeed, Ehrlich has shown convincingly that interest in non-Archimedean mathematics emerged in the 1870s and continued to grow in the hands of mathematicians such as Veronese, du Bois-Reymond, Levi-Civita, Hahn, Stolz, Hardy, and others. In the introduction to Ehrlich (2006), we read:

What is not so well known in these communities [historians and philosophers of mathematics], however, is that whereas most late nineteenth- and pre-Robinsonian twentieth-century mathematicians banished infinitesimals from the calculus, they by no means banished them from mathematics. Indeed, [. . .] between the early 1870s and the appearance of Abraham Robinson's work on non-standard analysis in 1961 there emerged a large, diverse, technically deep and philosophically pregnant body of consistent non-Archimedean mathematics of the (non-Cantorian) infinitely large and the infinitely small. Unlike non-standard analysis, which is primarily concerned with providing a treatment of the calculus making use of infinitesimals, most of the former work either emerged from the study of the rate of growth of real functions, or is concerned with geometry and the concepts of number and of magnitude or grew out of the natural evolution of such discussions. What may surprise many historians and philosophers even more than the existence of these two bodies of literature is that the latter such body contains constructions of systems of finite, infinite and infinitesimal numbers that are not only sophisticated

---

[4] The literature on non-standard analysis is very rich. See Dauben (1995) for a biography of Robinson with special emphasis on non-standard analysis; Salanskis (1999) is an excellent discussion in French of the mathematical and philosophical issues surrounding non-standard analysis. See also Goldblatt (1998) for a recent formal introduction and Cutland, di Nasso, and Ross (2006) for recent mathematical developments. The reader is referred to the extensive bibliographies contained in those volumes for further references. An interesting alternative to non-standard analysis, which allows for a development of substantive parts of mathematics goes under the name of (smooth) infinitesimal analysis. Smooth infinitesimal analysis differs from both ordinary and non-standard analysis by allowing nilpotent infinitesimals, namely linelets dx such that $dx \neq 0$ but $dx \cdot dx = 0$. The best exposition of the topic is Bell (1998).

[5] The interconnection with many important related topics such as Conway's surreal numbers and other alternative approaches to the construction of the real numbers, such as smooth infinitesimal analysis mentioned in the previous note, cannot be properly addressed here. See Salanskis and Sinaceur (1992), Ehrlich (1994), Schuster, Berger, and Osswald (2001), and Ehrlich (2012).

by contemporary mathematical standards but which are rich enough to embrace the corresponding number systems employed in non-standard analysis. (Ehrlich 2006, p. 3)

Once again, it would be out of place in this introduction to attempt even a small survey of the mathematical theories mentioned above. But the reader who will pursue those leads will come to appreciate the rich developments that eventually merge in the theories that constitute the contents of the third chapter of the book and that also led me to raise related issues on neo-logicism in the fourth chapter.

Chapter 3, in fact, focuses on a theory of counting with infinite sets that emerges from the work of Vieri Benci, and was developed with his collaborators, Marco Forti and Mauro Di Nasso.[6] The interesting thing here is that the problem of counting infinite sets, namely giving a "size" to infinite sets, is one that set theory seemed to have solved once and for all with Cantor's theory of cardinal numbers.[7] According to Cantor, we compare sets with respect to "size" (cardinality) using the following criterion: two sets A and B have the same cardinal number if and only if there exists a one to one correspondence between A and B. Cantor's great achievement was to show that this criterion allows us to distinguish different sizes of infinity (say that of the natural numbers and that of the real numbers) and to show that there is an infinite scale of infinities. The fruitfulness of this definition in mathematics and the lack of mathematically worked-out alternatives seemed to make any other approach impossible. Indeed, Gödel went as far as to offer an argument for why a generalization of arithmetic from the finite to the infinite would inevitably lead to Cantor's theory of cardinals.

But things turn out to be historically and mathematically much more interesting. Benci, Di Nasso, and Forti developed theories of counting infinite sets that not only yield the ordinary counting on finite sets but also assign "numerosities" to infinite sets by preserving the part-whole principle, namely the principle according to which if a set A is strictly included in a set B, then the numerosity of A is strictly smaller than the numerosity of B. The system of numerosities they developed for countable sets, for instance, satisfies the same algebraic laws as ordinary arithmetic (and in this sense it is much better behaved than cardinal arithmetic). But what I would like to emphasize here is that the origin of this development is to be found in the attempt at finding more intuitive foundations for non-standard analysis. Here is then how a development in non-Archimedean mathematics has the potential to also make us revisit a central concept of Cantorian set theory. Of course, as I explain at length in Chapter 3, the two developments are perfectly consistent with one another. While according to Cantor's criterion—an equivalence

---

[6] Bibliographical references can be obtained from the references provided in Chapter 3.
[7] For an introduction to Cantor's life and work see Dauben (1990) and Purkert (1987).

relation based on one to one correspondence—the even numbers and the natural numbers have the same cardinality, according to the theory of numerosities, the numerosity of the set of even numbers is strictly less than the numerosity of the natural numbers. The equivalence relation that leads to numerosities can even be seen as a refinement of the Cantorian relation yielding cardinalities. But the existence of a mathematical alternative to capturing the notion of "size" for sets leads to interesting mathematical, historical, and philosophical problems.

Mathematically, it opens the way to the search for new alternatives to the standard theory and to study the interaction of the new theories with other core mathematical areas, such as number theory and probability theory. Historically, it allows us to revisit the history of infinity with more sensitivity for those attempts that had been made to assign "sizes" to infinite sets while preserving the part–whole principle (this is analogous to the possibilities offered by the success of non-standard analysis in revisiting previous non-Archimedean developments). Finally, philosophically it allows us to discuss, as I do in Chapter 3, various philosophical claims that have been made in connection to Cantor's theory of cardinal numbers (e.g. Gödel's inevitability claim and Kitcher's discussion of rational transitions between mathematical practices).[8]

Abstraction and Infinity

Yet another area of great philosophical concern that has benefited from the developments discussed above is neo-logicism. What follows will set in context the emergence of Chapter 4 and bring to full fore the interaction between abstraction and infinity.

Neo-logicism is an attempt to revive Frege's program by claiming that important parts of mathematics, such as second-order arithmetic, can be shown to be analytic. The claim rests on a logico-mathematical theorem and a cluster of philosophical argumentations. The theorem is called *Frege's theorem*, namely that second-order logic with a single additional axiom, known as HP (Hume's Principle), deductively implies (modulo some appropriate definitions) the ordinary axioms for second-order arithmetic. The cluster of philosophical claims is related to the status (logical and epistemic) of HP. In the Fregean context the second-order systems have variables for concepts and objects (individuals). In addition, there is a functional symbol # that denotes a function that when applied to concepts yield objects as values. The intuitive meaning of # is as an operator that

---

[8] Recent developments focus on whether numerosities can be accepted as a theory of counting and on the foundations of probability theory (see Parker 2013 and Benci, Horsten, and Wenmackers 2013).

when applied to concepts yields an object corresponding to the cardinal number of the objects falling under the concept (numbers are thus construed as objects). Hume's Principle has the following form:

HP: $(\forall B)(\forall C)\,[\#x{:}(Bx) = \#x{:}(Cx) \leftrightarrow B \approx C]$

where $B \approx C$ is short-hand for one of the many equivalent formulas of pure second-order logic expressing that "there is a one–one correlation between the objects falling under B and those falling under C". As remarked, the right-hand side of the equivalence can be stated in pure second-order logic. The left-hand side gives a condition of numerical (cardinal) identity for concepts. The concepts B and C have the same (cardinal) number just in case there is a one–one correlation between the objects falling under them. In the ordinary framework of second-order logic # is a total operation on concepts and thus all concepts have a cardinality. We have '$\#x{:}(x \neq x)$' which will have to denote the cardinality of a concept having empty extension (for this reason Frege uses the concept $x \neq x$ to define '0' as '$\#x{:}(x \neq x)$') and # might also be used to define new concepts so that it makes perfect sense to have a term such as '$\#y{:}(\#x{:}(x \neq x) = y)$' which according to Frege's definitions will denote the number 1. Notice that there will also be an object corresponding to '$\#x{:}(x = x)$' which will turn out to denote an infinite number (since there are at least countably many objects obtainable in the system, namely the natural numbers). While in Cantor's theory there is no universal set (say a set satisfying the condition $x = x$) and thus no cardinal number of "the universe", it is obvious that Frege's criterion for assigning numbers to concepts agrees in essence with Cantor's cardinality criterion.[9]

In his book *Frege's Theorem* (2011), Richard Heck connected my 2009 article (Mancosu 2009, which forms the core of Chapter 3) on infinity and numerosities with neo-logicism and argued that the conceptual possibilities opened up by the theories of numerosities invited a line of argument against the analyticity of HP. It was by reflecting on Heck's argument that I was led to the idea that perhaps these new developments should also be used to inquire whether HP was after all the only principle from which second-order arithmetic could be derived. This led me to two questions. The first historical: were there any alternatives to Hume's Principle that had been considered by Frege's contemporaries? The answer turned out to be yes; Bolzano, Schröder, and Peano, for instance, favored principles of number assignment that diverge from HP on infinite concepts (i.e. concepts with an infinite extension). The second question was mathematical: would it be possible to formulate abstraction principles that behaved like HP on finite

---

[9] There are of course ways to block the totality of the # operator. I discuss them in Chapter 4.

concepts but diverged in their assignments to numbers for infinite concepts? Using a powerful result implicit in Heck's previous work the answer turned out to be yes; in fact there are infinitely many such principles.[10] All these principles are "good companions" of HP in the sense that they can be seen to have the same epistemic virtues as HP (consistency etc.), they are compatible with HP, and the axioms of second-order arithmetic can be derived from them. On the basis of this result I formulate a "good company" objection and investigate how different stripes of neo-logicism would respond to it. This also led to revisiting Heck's argument—based on numerosities—against the analyticity of HP and related philosophical issues (such as cross-sortal identification of abstracta).

Here then is an excellent case of how a recent mathematical development on infinity has yielded high dividends in the philosophical discussion of issues whose understanding had hitherto been severely constrained by the lack of mathematical alternatives. It is my hope that this might only be the beginning of a whole new set of investigations on infinity, abstraction, and their philosophical relevance.

One final word about editorial decisions. Given that many of the sources I quote are difficult to find, I have decided in most cases to provide in footnotes the original quotations in German, French, Latin, and Italian translated in the main text. When standard translations were already available, and providing the original text seemed unnecessary, I have not done so. In some cases the page numbers given next to the cited article refer to a more easily available reprint of the original. For instance, Weyl (1910, p. 303) refers to p. 303 of the reprint of Weyl (1910) in his collected works Weyl (1968). The reader will find all the necessary information about the original publication and the reprint in the bibliography so that no confusions should ensue (one only needs to check the range of page numbers in the two publications). With the exception of Chapter 3, which originates from an article published in 2009, the book originates from work done since 2013. In the case of Chapter 3, I have added where relevant a few recent additional references that have come to my attention since the original publication of Mancosu (2009). There is a small overlap in the historical parts of Chapters 3 and 4 connected to the discussion of Bolzano and Cantor. I did not eliminate the overlap because doing so would have negatively impacted the narrative of each individual chapter.

---

[10] However, it is not possible, as I show in Chapter 4, to formulate an abstraction principle acceptable to the neo-logicist that preserves the part-whole principle on the entire universe of concepts.

Acknowledgements

I have benefited enormously from conversations with the following colleagues who have been generous with comments and suggestions either for this book or for some of the articles on which it is based. They are listed in alphabetical order: Fabio Acerbi, Aldo Antonelli, Andy Arana, Jeremy Avigad, Luca Bellotti, Francesca Boccuni, Paola Cantù, Vieri Benci, Nico Bertoloni Meli, Patricia Blanchette, Davide Bondoni, Paolo Bussotti, Roy Cook, Leo Corry, Nadine de Courtenay, Ciro de Florio, Vincenzo de Risi, Mic Detlefsen, Jean Dhombres, Philip Ebert, Solomon Feferman, José Ferreirós, Dominique Flament, Marco Forti, Craig Fraser, Daniel Fremont, Paolo Freguglia, Sergio Galvan, Marcus Giaquinto, Alessandro Giordani, Ivor Grattan-Guinness, Jeremy Gray, Bob Hale, Michael Hallett, Richard Heck, Wes Holliday, Jim Hutchinson, Daniel Isaacson, Jonathan Israel, Akihiro Kanamori, Eberhard Knobloch, Richard Lawrence, Hannes Leitgeb, Neil Lewis, Godehard Link, Jesper Lützen, John MacFarlane, Sébastien Maronne, Robert May, Bernardo Mota, Massimo Mugnai, Cecilia Panti, Marco Panza, André Pétry, Nick Ramsey, Clara Silvia Roero, Marcus Rossberg, Thomas Ryckman, Frank Thomas Sautter, Jan Sebestik, Stewart Shapiro, Hourya Sinaceur, Jamie Tappenden, Justin Vlasits, James Walsh, Sean Walsh, Crispin Wright.

I am greatly indebted to Sophie Dandelet for the preparation of the text in LaTeX.

I would like to thank the Institute for Advanced Study (Princeton) and the Guggenheim Foundation for their generous support in 2008–2009. Most of the book was written in Munich in 2014 while I was a visiting professor at the Munich Center for Mathematical Philosophy at the Ludwig-Maximilians-Universität, Munich. I thank the UC Berkeley–LMU–Munich exchange for supporting my work and Hannes Leitgeb for having been a most generous host.

Finally, many thanks to Peter Momtchiloff for his support and, last but not least, to my wife Elena whose patience, if not infinite, is certainly boundless.

The first chapter of the book has not been published before. The other chapters originate, fully or partially, from the following publications:

"Measuring the size of infinite collections of natural numbers: Was Cantor's theory of infinite number inevitable?", *The Review of Symbolic Logic*, 2, 2009, pp. 612–646. [Mancosu 2009]

"*Grundlagen*, Section 64: Frege's discussion of definitions by abstraction in historical context", *History and Philosophy of Logic*, 36, 2015, pp. 62–89. [Mancosu 2015a]

"In Good Company? On Hume's Principle and the assignment of numbers to infinite concepts", *The Review of Symbolic Logic*, 8, issue 2, 2015, pp. 370–410. [Mancosu 2015b] (A previous version in French appeared in P. Mancosu, *Infini, Logique, Géométrie*, Vrin, Paris, 2015, pp. 73–122 [Mancosu 2015c])

"Definitions by abstraction in the Peano school", in C. de Florio and A. Giordani, eds., *From Arithmetic to Metaphysics. A Path through Philosophical Logic. Studies in honor of Sergio Galvan*, Berlin: de Gruyter.

I am thankful to Cambridge University Press for permission to use the articles I published in *The Review of Symbolic Logic* in 2009 and 2015.

1

The mathematical practice of definitions by abstraction from Euclid to Frege (and beyond)

1.1 Introduction

In section 63 of the *Grundlagen*, Frege introduces the possibility of defining identity between numbers in terms of one–one correspondence and mentions that this type of definition "seems in recent years to have gained widespread acceptance among mathematicians. But it raises at once certain logical doubts and difficulties, which ought not to be passed over without examination."[1] The first set of issues raised by Frege concerns the employment of the concept of identity. Since the concept of identity is more general than the concept of numerical identity one would expect the definition of numerical identity to use both the concept of number and the general concept of identity. But the definition Frege is envisaging cannot make use of the concept of number since what is being attempted is to define the latter in terms of numerical identity.[2] The process and the accompanying qualms are described as follows:

---

[1] The problems related to accepting one–one correspondence as the key property for defining the equality of numbers are addressed in Chapter 4.

[2] From this point of view, it is doubtful whether Schröder's and Kossak's approaches, which are mentioned by Frege alongside that of Cantor, are correctly interpreted as described here by Frege. Indeed, while both use one–one correspondence to define the notion 'sind in gleicher Anzahl' in Schröder and 'gleiche Zahlen' in Kossak, it is not the case that they are implicitly defining the notion of number in the way described by Frege. Kossak has an explicit definition of Zahl as "die zusammengesetzte Vorstellung von Eins und Eins usw." (1872, p. 16) And Schröder 1873 says: "Eine natürliche Zahl ist eine Summe von Einern" (1873, p. 5). Similar appeals to one–one correspondence that are not meant as a definition of number are found in lectures by Weierstrass, say the 1878 introductory lectures on analytic functions (Weierstrass 1988) and in Kronecker (see Kronecker 1887 and 1891 [Boniface and Schappacher 2001]). Recently Sean Walsh has investigated how these definitions in Weierstrass and Kronecker relate to the program of arithmetization of analysis. Our investigations were carried out independently and we in fact focus on very different issues.

We thus do not intend to define equality specially for this case, but by means of the concept of equality, taken as already known, to obtain that which is to be regarded as being equal. Admittedly, this seems to be a very unusual kind of definition, which has certainly not yet received sufficient attention from logicians; but that it is not unheard of may be shown by a few examples. (Beaney 1997, p. 110)

Section 64 then provides examples of such definitions from mathematical practice but without providing specific references to the mathematical literature. The first is a definition of identity of directions between lines based on the relation of parallelism.

The judgement 'Line *a* is parallel to line *b*', in symbols: *a//b*, can be construed as an equation. If we do this, we obtain the concept of direction and say: 'The direction of line *a* is equal to the direction of line *b*'. We thus replace the symbol // by the more general =, by distributing the particular content of the former to *a* and *b*. We split up the content in a different way from the original way and thereby acquire a new concept. (Beaney 1997, p. 111)

A related way of proceeding can be followed in defining when two planes have the same orientation [*Stellung*][3] and the identity of shapes [*Gestalt*] of two triangles can be defined in terms of similarity. A further identity, based on collinearity, is also offered. Given Frege's training as a geometer, it is perhaps not surprising that his examples are taken from this area of knowledge (I shall return to this issue later). However, his hesitation concerning the extent of previous use of such definitions ("it is not unheard of") conveys the impression that before Frege such definitions were quite exceptional.[4]

   The peculiar definitions Frege is talking about were soon to be called definitions by abstraction. The terminology is not Fregean and Frege uses the expression only with reference to Russell's use of it, in a letter to Russell.[5] In the recent literature on neo-logicism such definitions are also referred to, on account of their axiomatic use, as abstraction principles. Yet, despite a massive amount of philosophical and technical work on such principles, the contemporary analytic literature seems to have lost track of the rich mathematical practice from which the use of such definitions originated and of the extended logical reflection that was bestowed upon them in the late nineteenth and early twentieth centuries. One of the goals of this chapter is to show the widespread use of definitions by

---

   [3] On the translation of *Stellung* see the first footnote in Chapter 2.

   [4] Much of the secondary literature seems to agree. For instance, Angelelli says: "Frege regarded this procedure if not as his own invention at least as a method in which he was pioneering." (1984, p. 463).

   [5] On Frege on abstraction see Angelelli (1984). Frege must have been familiar with the expression 'définition par abstraction' at least since the time he received Peano's letter dated 14.10.1896 (Frege 1976, p. 192). He uses the expression tentatively in a letter to Russell, dated 28.7.1902: "Vielleicht nennen Sie das "Définition par abstraction" " (Frege 1976, p. 224).

abstraction in mathematical practice before Frege's discussion in the *Grundlagen*. A welcome offshoot of the investigation, to be developed in Chapter 2, is a better historical account of the sources that are likely to have influenced Frege in writing *Grundlagen* §64. Furthermore, in this chapter, I will also extend the historical discussion of definitions by abstraction to include some post-Fregean developments, for instance Hausdorff's and Weyl's accounts of definition by abstraction developed in publications dating from around 1910. I should point out from the outset that I will not here be concerned with another type of abstraction process, which is found in Cantor and finds a logical source in Überweg (1874), section 51, with roots going back to Locke, Berkeley, and Hume.[6] Yet, Cantor's definition of cardinal numbers is a definition by abstraction also in the sense relevant to this chapter, as we shall discuss.

1.2 Equivalence relations, invariants, and definitions by abstraction

The terminology 'définition par abstraction' was coined in 1894 in the Peano school although all the relevant concepts were already in place in Peano 1888 and in earlier writers. We will see in Chapter 2 the central role of the Peano school in the discussion concerning the logical status of such definitions and its impact on Russell.

But what exactly is a definition by abstraction? And what connection does it have to abstraction? Although these questions are obvious, the answers turn out to be less straightforward, in part because of the shifts this notion underwent as people theorized about it. As a first approximation, we can say that a definition by abstraction takes its start from an equivalence relation $\sim$ over a class of entities (let us call it Dom for domain). The equivalence relation can be thought of as capturing a common feature, a similarity, among the entities standing in the $\sim$ relation. One then ignores any individual features of the objects except whether they stand in the $\sim$ relation thereby 'abstracting' from any other feature distinguishing the elements so related. Consequently, the objects related by the equivalence relation are partitioned into mutually exclusive and jointly exhaustive classes in such a way that the objects related by $\sim$ belong to one and only one class. Let Dom be a given domain and $\sim$ a binary equivalence relation on it.[7] The definition by abstraction

---

[6] For a contemporary discussion of Cantorian abstraction and a defense of it, see Fine (1998); for Frege's rejection of Cantorian abstraction see Frege (1976, pp. 76–80) and the articles by Angelelli quoted in the bibliography. For Cantor's criticism of Frege see Ebert and Rossberg (2009).

[7] Generalizing to $2n$-ary relations is unproblematic.

(or the principle, when it is so used) lays down a condition to the effect that this identification of the elements is captured by a functional relation F so that the following is satisfied:

For all a, b in Dom, $F(a) = F(b)$ iff $a \sim b$.

The function F plays the role of abstracting from the peculiar nature of the entities a and b and regards them as associated to the unique value $F(a)(= F(b))$ on account of the fact that a and b are related by the equivalence relation $\sim$ and thus, from the point of view of $\sim$, they are considered as indistinguishable.

Even before the Peano school dubbed this type of definition 'by abstraction', other authors, including Kronecker and Helmholtz, had already gotten to the essence of the situation. I will in fact take my start from what Kronecker says in 1889–1890 in his article 'Zur Theorie der elliptischen Functionen' (Kronecker 1889–1890; see also Kronecker 1891):

> With the happy expression "Invariants" chosen by Mr. Sylvester, and quite appropriate to the meaning of the matter, one originally denotes only rational functions of the coefficients of forms that remain unchanged under certain linear transformations of the variables of the forms. But the same expression has since then also been extended to some other entities [*Bildungen*] that remain unchanged under transformation. This multiple applicability of the concept of invariants rests upon the fact that it belongs to a much more general and abstract realm of ideas. In fact, when the concept of invariants is separated from the direct formal relation to a process of transformation and it is tied rather to the general concept of *equivalence*, then the concept of invariants reaches the most general realm of thought. For, every abstraction,—say an abstraction from certain differences that are presented by a number of objects,—states an equivalence and the concept originating from the abstraction, for instance the concept of a species, represents the "invariant of the equivalence".[8]

I will postpone to section 1.4.1 a more precise discussion of the mathematical number-theoretic background, for in this section I am merely concerned with

---

[8] "Mit dem von Hrn. *Sylvester* glücklich gewählten, sinnentsprechenden Ausdruck "Invarianten" sind zwar ursprünglich nur rationale Functionen der Coefficienten von Formen bezeichnet worden, welche bei gewissen linearen Transformationen der Variabeln der Formen ungeändert bleiben, aber derselbe Ausdruck ist seitdem schon auf mancherlei andere, bei Transformationen ungeändert bleibende Bildungen übertragen worden. Diese vielfache Anwendbarkeit des Invariantenbegriffs beruht darauf, dass derselbe einer weit allgemeineren abstracteren Ideen-sphaere angehört. In der That wird der Invariantenbegriff, wenn er von den unmittelbaren formalen Beziehung auf ein Transformationsverfahren losgelöst und vielmehr an den allgemeinen Aequivalenzbegriff geknüpft wird, in die allgemeinste Denksphaere erhoben. Denn jede Abstraction, z.B. die von gewissen Verschiedenheiten, welche eine Anzahl von Objecten darbietet, statuirt eine Aequivalenz, und der aus der Abstraction hervorgehende Begriff, z.B. ein Gattungsbegriff, bildet die "Invariante der Aequivalenz"." (Kronecker 1889–90, pp. 58–59).

giving a description by broad strokes of the context that led to the mathematical techniques that finds expression in quotes such as the aforementioned one.

Peano (1888)[9] also refers in this connection to a process of abstraction (but not to invariance); but we will see that the connection to abstraction is not original with Kronecker or Peano but goes back at least to Grassmann. I will come back to the connection between the notion of invariance and equivalence relations because it will shed light on Weyl's description of the process.[10] What Kronecker is giving us here is the (mathematical) description of a process that abstracts from specific peculiarities of the entities under consideration, thereby classifying them according to a more 'abstract' concept. For instance we might abstract from the individual properties of Tom, Dick and Harry, and all other persons, to consider them only as human beings. This is tantamount to introducing an equivalence relation that relates all human beings to one another. Having described the general idea, Kronecker then proceeds to highlight the importance of this process for any scientific activity:

Every scientific task aims at the determination of equivalences and at the discovery of their invariants and for it the following line of poetry holds: "the wise one seeks a stable pole amid the flight of phenomena".[11,12]

A similar, even more emphatic, description is given in Kronecker's lectures on the concept of number in 1891.[13]

---

[9] "Ogni uguaglianza fra gli enti d'un sistema, diversa dall'identità, equivale all'identità tra gli enti che si ottengono da quelli del sistema dato astraendo da tutte e sole quelle proprietà che distinguono un ente dai suoi uguali. Così l'eguaglianza 'il segmento AB è sovrapponibile ad $A'B'$' equivale all'identità tra gli enti che si ottengono da ogni segmento astraendo da tutte le proprietà che lo distinguono da quelli con cui è sovrapponibile. L'ente che risulta da questa astrazione vien chiamato *grandezza* del segmento; l'eguaglianza precedente equivale quindi all'identità delle grandezze dei due segmenti." (Peano 1888, p. 153).

[10] The secondary literature on Weyl on this topic (see Angelelli 1984 and Christopolou 2014) emphasizes the notion of invariance but fails to point out its roots in previous work by Weyl (e.g. Weyl 1910) and in the number-theoretic tradition that inspired him.

[11] The citation is from Schiller's *Der Spaziergang* (1795) and it was also Helmholtz's favorite citation (see Meulders 2010, p. 71).

[12] "Jede wissenschaftliche Forderung geht darauf aus, Aequivalenzen festzustellen und deren Invarianten zu ermitteln, und für jede gilt das Dichterwort: "der Weise sucht den ruhenden Pol in der Erscheinungen Flucht"." (Kronecker 1889–90, p. 59).

[13] "Und nun noch ein Wort über die Bedeutung dieser Begriffe. Das Aufsuchen der Invarianten ist eine schöne, ja sogar die schönste Aufgabe der Mathematik; aber noch mehr, es ist sogar ihre einzige Aufgabe. Und auch damit ist es noch nicht genug: es ist die einzige Aufgabe aller Wissenschaften überhaupt. Das Setzen von äquivalenzen und das Aufsuchen ihrer Invarianten, das ist seinerseits die Invariante jeder Forschung und jeder geistigen Arbeit. Und um nur eins von den unerschöpflich vielen als Beispiel hier anzuführen: ein jeder Begriff ist die Invariante der Individuen, welche unter ihm als sein Inhalt gedacht werden. Wollen wir aber diesen Begriff, welcher allüberall in der Sphäre menschlichen Denkens und Schaffens die unbeschränkte Herrschaft führt, in ein königliches Gewand kleiden, so können wir nur wieder die Worte Schillers zitieren, welcher die Forscherarbeit des Weisen

Kronecker's detailed description of the process uses n-tuples $(z_1, \ldots, z_n)$, which he calls systems, abbreviated as (z). The starting point is an equivalence relation on these systems satisfying the property:

If $(z) \sim (z')$ and $(z) \sim (z'')$, then $(z') \sim (z'')$[14]

and specifies that the systems related by $\sim$ can be united [*vereinigt*] into one and the same 'class'. I will now follow the wording of the 1891 lectures since they are more explicit about the description of the process.

Kronecker's interpretation of the equivalence relation is very much influenced by work in number theory, especially his work on binary quadratic forms. A binary quadratic form is an expression of the form[15] $ax^2 + 2bxy + cy^2$ and it can be identified as a triple of numbers (a, b, c). In the case of binary quadratic forms $n = 3$ and thus Kronecker sees the equivalence relation as capturing the idea that one triple can be reached, through an inferential or computational process, from another triple.[16] That means that starting from a representative of the equivalence class, identified as the characteristic element of the class, one should be able to reach all the other elements of the equivalence class. Conversely, starting from any binary quadratic form in any one class, one should be able to compute the representative to which that binary quadratic form belongs (see section 1.4.1 for further details). This leads to the following consideration:

We now come to the consideration of the invariants of equivalent systems and ask ourselves the question: are there functions of the elements that have the same value for the entire class of equivalent systems?[17]

The existence of such a function, adds Kronecker, is trivial if one uses the general definition of function.[18] However, Kronecker is interested in a finer classification

---

in den einzelnen Wissensgebieten schildert und dann zusammenfassend sagt: "Der Weise sucht den ruhenden Pol in der Erscheinungen Flucht". " (Kronecker 1891, pp. 235–236).

[14] In the 1891 lectures Kronecker shows explicitly how to get reflexivity and symmetry (although he has no specific terminology for these properties) out of this condition. He assumes that the relation is such that for any (z) there is a (z') such that $(z) \sim (z')$. More on the condition below.

[15] Kronecker in his lectures uses $ax^2 + bxy + cy^2$ as a representation for a generic quadratic form. I use $ax^2 + 2bxy + cy^2$ for consistency with Gauss's original formulation and with Dirichlet's lectures on number theory which I will follow in section 1.4.1.

[16] This effectiveness is not something that is required or emphasized in the usual set-theoretic treatment of these matters nowadays although they are obviously essential to applications in number theory.

[17] "Wir kommen jetzt zu der Betrachtung der Invarianten äquivalenter Systeme und stellen uns die Frage: giebt es Funktionen der Elemente, welche für die ganze Klasse äquivalenter Systeme denselben Wert haben?" (Kronecker 1891, p.237).

[18] Kronecker might be thinking here of using the equivalence class as the value of the function. As we will see he warns against this possible solution for it provides no computational information.

and points out that the type of invariant will depend on the nature of the process (namely, a function) yielding the elements of the canonical representative for each equivalence class. He classifies invariants as arithmetical, algebraic, and analytic:

One can reply: for each arbitrary equivalence there are always invariants in the above mentioned sense. This is indeed a trivial and completely uninformative truth, for it is the broad concept of function that puts us in a position to give an answer. Let us consider again the already mentioned equivalence for a system of three whole numbers, then we can conceive of the elements of an arbitrarily selected system as characteristic invariants. The elements are invariants because I can reach this system from any other system and they are the characteristic invariants because they represent a system that belongs to one definite class but to none of the other classes. In this typical case of quadratic forms [[which are identified as triples of numbers, PM]], one chooses as the characteristic invariants of the class the elements of that system that have the smallest value and calls, since Lagrange's time, the form represented by them the reduced form (*forme réduite*). From it one can of course generate the entire class of forms. I distinguish the invariants as arithmetical, algebraical, and analytical, according to the method through which they are derived from the elements of a system.[19]

Somewhat confusingly, but perfectly in line with a usage still pervasive in much contemporary mathematical practice, Kronecker in his work speaks of invariants both in reference to the elements that make up the reduced form (i.e. the triple (a, b, c) that is the reduced form for its equivalence class) as well as in reference to the function that generates the canonical representatives (our terminology), namely the function F that is such that $F(a', b', c') = (a, b, c)$, for any triple (a', b', c') in the equivalence class of the canonical representative (a, b, c).[20]

Another possibility is that he is referring to the choice of an arbitrary element from the equivalence class, rather than a canonical representative.

[19] "Man kann antworten: es giebt für jede beliebige äquivalenz stets Invarianten in dem angeführten Sinne. Dies ist eigentlich eine triviale, weil völlig nichtssagende Wahrheit, da nur der weitgreifende Begriff der Funktion uns ermöglicht, die Antwort zu geben. Nehmen wir wieder die genannte äquivalenz für Systeme von drei ganzen Zahlen, so können wir die Elemente eines beliebig herausgegriffenen Systems als charakteristische Invarianten begreifen. Die Elemente sind Invarianten, weil ich von jedem System zu diesem einen gelangen kann, und sie sind die charakteristischen Invarianten, weil sie ein System repräsentieren, welches nur der einen bestimmten Klasse, aber keiner anderen zugehört. Man wählt in diesem typischen Falle der quadratischen Formen die Elemente desjenigen Systems als die charakteristischen Invarianten der Klasse, welche den kleinsten Wert haben und nennt die daraus gebildete Form schon seit Lagrange die reduzierte Form (*forme réduite*). Aus derselben kann man sich natürlich die ganze Klasse von Formen herstellen. Ich unterscheide die Invarianten als arithmetische, algebraische und analytische Invarianten nach der Methode, durch welche sie aus den Elementen eines Systems hergeleitet werden." (Kronecker 1891, pp. 237–238).

[20] Commenting on the situation Kronecker raises the issue of invariance by applying the terminology of functions: "Giebt es Funktionen der Zahlen (a, b, c), welche so beschaffen sind, daß sie für sämtliche Systeme derselben Klasse identisch sind? Ist eine solche Funktion vorhanden, so legen wir ihr die von Sylvester so außerordentlich glücklich gewählte, obgleich von ihm nur für einen ganz speziellen Fall gebrauchte Bezeichnung 'Invariante'." (Kronecker 1891, pp. 234–235) See also this

The characteristic invariants are thus those elements that are canonical representatives of the class and, by extension, one calls them arithmetical, algebraical or analytical according to the nature of the process (i.e. the invariant function) that allows one to move from the elements of a triple to any other elements of the class.

Let's postpone further details to section 1.4.1 and let us take stock. If one abstracts from the niceties of calibrating the exact nature of the invariant function, namely, whether it be arithmetical, algebraic or analytical, which is however essential for Kronecker,[21] we have here in essence a description of the method that, given an equivalence relation over a domain of objects, from elements of an equivalence class yields its representative. The process, expressed functionally, satisfies the condition:

For all a, b in Dom, $F(a) = F(b)$ iff $a \sim b$.

Remarkable in Kronecker is also the emphasis, as I have already pointed out, on how the process corresponds to the formation of concepts by abstraction. The reason why I started from this exposition of Kronecker can already be introduced now. For, while Kronecker speaks of a process of abstraction and concept formation, it can be argued that the specific cases of number theory he is mostly concerned with neither amount to the introduction of concepts by abstraction (except in an etiolated sense), nor to new abstract objects, but rather define functions whose values belong to the original domain on which the equivalence relation is defined. Thus, despite the fact that what we normally call definitions by abstraction fall under the general description given above, it is not prima facie clear that all procedures leading from equivalence classes to their representatives end up being relevant definitions by abstraction. I will say more

account of Kronecker's work by Netto: "Jede Abstraction, z. B. die von gewissen Verschiedenheiten, welche eine Anzahl von Objecten darbietet, statuirt eine Aequivalenz; alle Objecte, die einander bis auf jene Verschiedenheiten gleichen, gehören zu einer Aequivalenzclasse, sind unter einander aequivalent, und der aus der Abstraction hervorgehende Begriff bildet die "Invariante der Aequivalenz." Die besondere Aufgabe der Mathematik ist es, diese Invarianten in Gestalt identischer Gleichungen zu formuliren. Wenn also $z'_1, z'_2, \ldots, z'_n$ beliebige Objecte sind, und $z'_\alpha, z''_\alpha, z'''_\alpha, \ldots$, einander aequivalent (für $\alpha = 1, 2, \ldots, n$), dann müssen die für die Aequivalenz charakteristischen, eindeutigen Functionen I_k gefunden werden, welche

$$I_k(z'_1, z'_2, \ldots, z'_n) = I_k(z''_1, z''_2, \ldots, z''_n) = I_k(z'''_1, z'''_2, \ldots, z'''_n) = \ldots$$

liefern. Die I_k müssen durch die Gleichungen einen vollständigen Ersatz des Aequivalenzbegriffes geben. Solche Invarianten heissen rationale, algebraische, arithmetische, analytische, je nachdem sie durch rationale, algebraische, arithmetische oder analytische Operationen aus den Elementen gebildet werden, und unter analytischen Operationen werden hierbei solche verstanden, bei denen der Limesbegriff zur Anwendung kommt." (Netto 1896, p. 246).

[21] This is one of the reasons on account of which Kronecker (1882) warns against the use of the equivalence class as the value of the function. See Kronecker (1895–1930, vol. II, p. 356).

about this later but let me for the moment explain briefly what I have in mind. If we think of a definition by abstraction as generating abstracta, namely abstract elements $F(a)$ for any element a in Dom, then the number theory case does not seem to yield abstract objects in any thick sense of the word simply because F outputs a value which is already contained in the original domain over which the function is defined and thus the value is just as concrete as any object in the original domain of objects we started with. The situation seems different when we introduce directions through a function satisfying $\mathrm{dir}(a) = \mathrm{dir}(b)$ iff $a//b$. In the latter case it is normal to treat (as does the neo-logicist literature) $\mathrm{dir}(a)$ as an abstract object which is not identified with a line given in the original domain of lines. But let's postpone this discussion for the moment.

The style of doing mathematics through equivalence relations and equivalence classes (or representatives thereof) is so familiar to us that we often forget how long it took for it to become a standard technique and for the terminology to get settled. Thus, I would like to add a few more comments on equivalence relations before we move on. The *terminus ad quem* of the process leading to a set-theoretic presentation of how to go from an equivalence relation to its equiv-alence classes can be identified with the treatment given in §5 (*Klasseneinteilung. Äquivalenzrelation*) of van der Waerden's *Abstrakte Algebra* (1930, p. 14). There, an equivalence relation is defined as one that satisfies reflexivity, symmetry and transitivity (with transitivity defined, as is now standard, as "if $a \sim b$ and $b \sim c$, then $a \sim c$"). In that section it is shown how from an equivalence relation one can construct, set-theoretically, the equivalence classes associated to the relation and how to recover an equivalence relation from jointly exhaustive and mutually exclusive classes.[22] The examples offered come from number theory (congruence mod n) and geometry (congruence). Van der Waerden was not the first to use the terminology of 'equivalence relations' (see also Hasse 1926, pp. 16–18, and Haupt 1929, pp. 2–6) but his textbook played a central role as a classic presentation of this standard mathematical technique.

Weyl 1910 also contains a description of definition by abstraction. Weyl takes his start from the relation of equinumerosity between two sets and then generalizes to an arbitrary equivalence relation.[23] This gives rise, says Weyl, to a definitional

---

[22] This is not original with van der Waerden and the technical details were already clear in the 1910s.

[23] "So wird denn zunächst das aus dem symbolischen Zählverfahren erwachsene Kriterium für die Gleichzahligkeit zweier Mengen, welches darin besteht, daß man die zu vergleichenden Mengen elementweise umkehrbar eindeutig auf einander abbilden kann, zur *Definition* der Gleichzahligkeit erhoben. Um daraus den Begriff der Anzahl selbst zu gewinnen, kann man sich eines auch sonst vielfach in der Mathematik benutzten Definitionsverfahrens bedienen, das seine psychologischen Wurzeln in dem *Abstraktions*prozeß hat. Die Gleichzahligkeit, so wie sie soeben erklärt wurde, ist

process that is often used in mathematics and that has its psychological roots in the abstraction process. He points out that in mathematical practice such definitions are found galore. The interesting thing here is that we are getting quite close to the use of the word 'equivalence relation' to single out relations whose properties and importance had already been fully grasped by then in mathematical practice and that the Peano school had already subjected to analysis. It is also interesting that Weyl does not define the properties of the equivalence relation as Hasse, Haupt and van der Waerden do but rather he uses reflexivity and the property that if $a \sim b$ and $b \sim c$ then $c \sim a$.[24] The Kroneckerian roots of Weyl's discussion, among others, are revealed also by his reference to 'invariance' (see Weyl 1910, p. 303; see also Weyl 1927). Two comments might be useful at this juncture.

The first is that it took more than a century for the word 'Äquivalenz' to be used for characterizing equivalence relations in general. Use of the word 'equivalence' appeared all over the place in number theory (Gauss, Dirichlet, Eisenstein, Dedekind etc.), algebra (see for instance Cauchy 1847 on complex numbers and Grunert's 'Theorie der Aequivalenzen' in 1857, Dedekind's early work on equivalent groups in the 1850s, and Weber's 1896 *Lehrbuch der Algebra*, vol. II), in logic for equivalent propositions, and, as mentioned, Cantor used it as a central concept in set theory (see also Hausdorff 1914, p. 25, and Fraenkel 1928, pp. 19–20).

The second point of interest is the specific way in which Kronecker and many others define the properties for being an equivalence relation by giving pride of place to the property 'if $a \sim b$ and $a \sim c$ then $b \sim c$'. While it is a trivial fact that the latter definition of equivalence (when using also an extra property such as reflexivity) and van der Waerden's definition using transitivity amount to the same notion (see Vailati 1892 and De Amicis 1892), historically the difference is important. This property used by Kronecker and others turns out to be more often used in Greek mathematics than our property of transitivity. It is in fact

---

eine Beziehung zwischen zwei Mengen vom *Charakter der Äquivalenz*, d. h. eine Beziehung welche die folgenden Eigenschaften besitzt:

$$a \sim a$$
Aus $a \sim b, b \sim c$ folgt $c \sim a$

Allemal nun, wenn eine solche Beziehung zwischen Objekten a, b,...erklärt ist, hält man es für möglich, jedem der Objekte a derart ein anderes Ding α zuzuweisen, daß zwei Objekten dann und nur dann dasselbe Ding zugewiesen erscheint, falls jene beiden Objekte im Sinne der Beziehung $\sim$ einander äquivalent sind. Beispiele in der Mathematik, wo ein solches Verfahren zur Definition eines neuen Operationsbereiches von Dingen α benutzt wird, ließen sich in Hülle und Fülle beibringen." (Weyl 1910, in Weyl 1968, vol. I, p. 303).

[24] The equivalence of this Weylian definition to the one given by Kronecker and to one given by van der Waerden is easy to prove.

captured by Euclid's common notion 1 and it is used most of the time as a criterion for introducing quantities in many mathematical areas up to the early twentieth century.[25] Indeed, it seems to have been the preferred property in much of nineteenth century number theory and in other areas such as geometry.[26]

Notwithstanding the above, it was in the Peano school that for the first time the three properties characterizing an equivalence relation were assigned a name. It was with Padoa 1908 that such relations acquired the characterizing name of 'relazione egualiforme' defined as a relation that satisfies reflexivity, symmetry, and transitivity (1908, p. 95).

1.3 Mathematical practice and definitions by abstraction in classical geometry

Peano was quite aware that the interest of definitions by abstraction rested on their widespread use in mathematical practice. Indeed, the Peano school repeatedly pointed out that the first definition by abstraction went back to Euclid and singled out the definition of 'same ratio' as a key example.

Euclid's definition is found in book V definition 5 of the Elements:

Magnitudes are said to be in the same ratio, the first to the second and the third to the fourth, when, if any equimultiples whatever be taken of the first and third, and any equimultiples whatever of the second and fourth, the former equimultiples alike exceed, are alike equal to, or alike fall short of the latter equimultiples respectively taken in corresponding order. (Euclid 1956, Heath translation, vol. 2, p. 114)

'Ratios' are here introduced not by defining them explicitly, but by providing a criterion of equality. One must remark that the definition by abstraction involved here requires a four place relation as opposed to the often cited ones (such as that yielding directions for lines) which require only binary relations. This is because we are defining what it means for the ratio of a and b ($a{:}b$) to be equal to the ratio

---

[25] See Acerbi 2009 also for connections to Aristotelian logic and evidence showing that the Greeks preferred giving demonstrations through the use of the property described by common notion 1 rather than using transitivity.

[26] For instance Gauss, Kummer, Dedekind, and Kronecker prove that congruency modulo n is an equivalence relation by proving the equivalent of Euclid's common notion 1 for the relation in question (same for more complicated equivalence relations) as opposed to giving separate arguments for symmetry and transitivity. By contrast, Dirichlet (1863, p. 34; 1879, p. 34) uses ordinary transitivity but on p. 140, when dealing with the proper equivalence of quadratic forms he uses Euclid's common notion 1. Appeal to the latter property seems to have been the norm also with mathematicians writing in areas other than number theory (Grassmann, Hamilton, and Helmholtz, for instance).

of c and d (c:d). The relation R(a, b, c, d) corresponding to the right-hand side of the Euclidean definition is an equivalence relation. [27]

It might be worthwhile to point out from the outset that, in contrast to the relative clarity of this example with ratios, when reading sources in classical mathematics one often finds situations in which one is naturally led to interpret the mathematical development in terms of abstracting from an equivalence relation to obtain the associated abstract objects (variously introduced as new sorts of ideal objects, as equivalence classes or as representative elements of the latter). However, the interpretation is not necessarily faithful to the mathematical practice that is being scrutinized. Thus, before moving on to discuss other examples of definitions by abstraction, it will be important to reflect on at least two factors that create problems for the historian fishing for clear cut examples of definition by abstraction in past mathematics.

The first type of problem emerges from the fact that the mathematical text might be ambiguous as to whether an equality between new abstract terms is being defined at all, or as to whether there is an explicit specification of the (equivalence) relation that is undergirding the introduction of the new equality. Consider the notion of equality of segments in Euclid. Starting from the congruence relation between segments, a contemporary mathematician might naturally introduce length using a definition by abstraction such as '$l(a) = l(b)$ iff a is congruent to b' (with the option of explicitly defining length by means of equivalence classes or other devices or simply accepting lengths as new entities, as Peano does). But Euclid does not do this and he simply says, in common notion 4, that two segments are equal if they 'coincide with one another' ('Things which coincide with one another are equal to one another'). Then in the midst of the proof of proposition I.4 we find the converse being implicitly used for segments ('if two segments are equal they coincide with one another'). Is the notion of equality of segments taken to be primitive or is it introduced by abstraction (for it is not defined explicitly)? If we exclude the former case then, if there is a definition by abstraction of equality of segments, it is at best implicit, for what we are originally given is not a definition introduced by an 'if and only if' (and a fortiori not a definition by abstraction). Moreover, there is no mention of the class of segments that have in common the property of being congruent (which in a more contemporary setting could be used to define $l(a)$, namely the length of a.).

---

[27] There is an obvious way to extend the terminology of equivalence relations to $2n$-ary relations. Of course, Euclid does not speak of equivalence relations and it is doubtful whether at this stage it was clear exactly what properties of the relation need to be in place for the definition to aspire to being successful.

The second problem concerns the fact that often it is unclear whether a definition of equality between the newly introduced terms is meant as a 'thick' way to define a new concept or whether the identity is simply based on a previous explicit definition of the concept under which the objects whose identity is being defined fall. We will encounter many such examples but a good classical instance here is the very definition of equality of ratio in the Euclidean tradition. When we look at the early modern interpretations of Euclidean proportion theory, as exhaustively studied by Enrico Giusti in *Euclides reformatus* (1993), we encounter two traditions. The first originates from Clavius and offers a philosophical definition of ratio, which is then applied by means of the Eudoxian theory of equality of ratios; the second, going back to Commandino, defines the equality of ratios directly without an explicit definition of ratio (either philosophical or mathematical). In the former case, the definition by abstraction is 'thin', for it rests on an explicit definition of ratio (it is indifferent at this stage whether the definition is philosophical or mathematical) from which one then infers the relevant biconditional. In the latter case, the definition is 'thick', in that the notion of ratio has no other definitional introduction than by means of the identity appearing in the left-hand side of the relevant abstraction principle. In concrete cases, it is not always possible to judge whether a mathematician is working with a 'thin' or a 'thick' definition by abstraction.[28] Let me illustrate this point with another candidate for definition by abstraction, namely Desargues's definitions of 'ordinance of straight lines' and 'ordinance of planes' in the *Brouillon Projet* of 1636, one of the classical sources of projective geometry. Let us focus on 'ordinance of straight lines'. Desargues considers straight lines as infinite in both directions. Then he says:

To convey that several straight lines are either parallel to one another or are directed towards the same point we say that these straight lines belong to the same ordinance, which will indicate that in the one case as well as in the other it is as if they all converged to the same place.

The place to which several lines are thus taken to converge, in the one case as well as in the other, we call the *butt* of the ordinance of the lines.

To convey that we are considering the case in which several lines are parallel to one another we often say that the lines belong to the same ordinance, whose butt is at an infinite distance along each of them in both directions.

---

[28] In the *Grundlagen*, Frege first explores the possibility of defining numerical identity through a 'thick' definition by abstraction and, having found the outcome unsatisfactory, he gives an explicit definition of 'the number of a concept C' thereby obtaining the biconditional corresponding to the definition by abstraction as a theorem based on the explicit definition of 'number of concept C' as an extension. I call the latter a 'thin' definition by abstraction in that the biconditional characterizing the definition by abstraction is obtained as a theorem by means of an explicit definition of the concept appearing on the left-hand side of the biconditional.

To convey that we are considering the case in which all the lines are directed to the same point we say that all the lines belong to the same ordinance, whose butt is at a finite distance along each of them.

Thus any two lines in the same plane belong to the same ordinance, whose butt is at a finite or infinite distance. (Field and Gray 1987, p. 70)

In a similar way, Desargues defines the ordinance of planes and the 'axle', defined as the line of intersection of two planes or the place at infinity where parallel planes "are imagined to converge".

Let us consider the definition of 'ordinance of straight lines' in more detail. In Euclidean geometry every two non-parallel lines on the same plane meet at a point. Desargues generalizes this also to the case of parallel lines (we are thus in a projective context). We can think of two parallel lines meeting at a place (what we would now call a point at infinity although Desargues does not quite say that). Both in the case of intersecting lines in ordinary geometry and in that of parallel lines we call *butt* the point of intersection (Desargues says 'place' but I will use 'point' in my comments). Notice that the butt is intimately connected to the notion of 'direction'. However, 'direction' is not used substantively (as in 'the direction of b') but only relationally (b is directed towards a point C). For this reason, although I consider these passages as the forerunner of the type of abstraction principle on direction given by Frege (we will see later more plausible sources for Frege's appeal to the direction principle), I will not interpret such talk of directions as containing an abstraction principle. But the concept of *butt* is intimately related to that of direction and the way it is introduced for parallel lines brings us very close to a definition by abstraction.

Let us now consider, of the two cases considered by Desargues, the case in which we focus on the relation of parallelism, denoted by $a//b$. This is an equivalence relation over all lines on a given plane and we can thus introduce a function, call it butt*(a)[29] that gives us an ideal element corresponding to a point at infinity towards which all the lines parallel to a are directed. Desargues's definition of *butt* as "the place to which several lines are thus taken to converge" in the case of parallel lines is not defined explicitly as a point of intersection but must rather be introduced through the relation of parallelism. Thus, we end up with

butt*(a) = butt*(b) iff $a//b$.

---

[29] Please note that the * has been added to 'butt' to alert the reader that butt*(x) does not correspond to the general definition of *butt* as found in Desargues but only to a restriction that excludes the case of lines that intersect at a finite point.

In Desargues, butt*(a) stands for an ideal element that is neither identified with a previously given ordinary point, nor with the line a, nor with the class of all lines parallel to a. Despite the fact that Desargues does not explicitly restate the equivalence relation into an equality, and provides what looks like an attempt at an explicit definition in terms of 'place', I think a definition by abstraction comes very close to what he is actually doing.

That lines a and b belong to the same ordinance can, in the case of parallel lines, now be expressed by

$$\text{BSO}(a, b) \text{ iff butt}^*(a) = \text{butt}^*(b).$$

It should be evident that, despite my carefulness above, the way is now open to interpret the notion of butt*(a) as that of direction. That is the way in which later projective geometers will present this Desarguesian intuition.

Classical mathematics up to the end of the seventeenth century makes only spare use of definition by abstraction and most cases I have looked at, for instance Cavalieri's notion of 'all the squares of a plane figure', suffer from the definitional ambiguity we also found in Desargues.

Whatever one might think of the best way to interpret what Desargues is saying, the ambiguity of the text is an invitation to make the following remark. Until the treatment of definitions by abstraction that we will encounter in Grassmann, most of the examples found in mathematical practice are ambiguous between a relational and a substantival usage of the biconditional corresponding to a definition by abstraction. Let me clarify this with the notion of direction. If we say 'lines a and b have the same direction if and only if a is parallel to b' the claim can be given a relational reading or a substantival reading. According to the first, we are introducing a relation between a and b, call it $\text{HSD}(a, b)$, which can be taken as an unanalyzable short-hand for 'a is parallel to b': No new terms (or corresponding entities) are introduced in this way. In the substantival reading 'lines a and b have the same direction' is read as 'the direction of a equals the direction of b' and this introduces terms for directions. (more formally, $\text{dir}(a) = \text{dir}(b)$). As I mentioned, this ambiguity is often noticeable in practice but in Grassmann, and later Frege, the formulation of abstraction principles will be unambiguously substantival (see section 1.4.3).[30]

---

[30] The reader familiar with the neo-logicist literature will not fail to see in this ambiguity the root of two possible positions which will lead to a theoretical debate on the meaning of abstraction principles presented at length in Wright (1983, p. 68), namely an austere one that eschews the introduction of new entities and insists on reading the left-hand side of the biconditional, despite its apparent syntactical complexity, as an unanalyzable predicate, and a platonist reading which sees abstraction principles as introducing new sorts of expressions which function syntactically and semantically as terms denoting abstract objects. See also Heck (2011, Chapters 8 and 9).

Leibniz seems to have had a clear understanding of what is peculiar to a ('thick') definition by abstraction. In his fifth reply to Clarke, Leibniz had declined to give a direct definition of 'place' and had offered a definition of 'same place'. He then made the following insightful comment:

To conclude: I have here done much like Euclid, who not being able to make his readers well understand what *ratio* is absolutely in the sense of geometricians; defines what are the *same ratios*. Thus, in like manner, in order to explain what *place* is, I have been content to define what is the *same place*. (Leibniz 1955, p. 71)

Of course, Leibniz does not spell out what formal properties the relation on the right-hand side must satisfy but he sees clearly that a definition by abstraction allows us to talk about certain entities by defining what their equality means even if we are not in a position to provide an explicit definition of the terms flanking the equality. In addition, he did consider definitions by abstraction to be epistemologically inferior to explicit definition,[31] and this is a theme that will recur in the history of definitions by abstraction.

I will conclude this section by briefly mentioning another interesting case in which definition by abstraction seems to underlie a significant mathematical development but in which the definition by abstraction is only implicit. I am referring to Euler's introduction of general equations for curves in his *Introductio in Analysin Infinitorum* (Euler 1748, 1990). In volume 2, section 2, of that work, Euler is concerned with the fact that a geometrical curve that can be described by a certain equation under one coordinate system will receive a different description under a different coordinate system (rectangular or oblique). Yet, the curve, intensionally defined through these different modes of presentation remains extensionally the same. Indeed, when defining two coordinate axes we have the freedom to choose the horizontal line (corresponding to the variable x) and on it a point which will determine the y axis (if we use oblique axes we are also free to choose the angle between the y axis and the x axis). Euler observes:

Since [the origin and the axis] may be varied in an indefinite number of ways and even for the same curved line innumerable equations can be given, on this account from the difference of the equations one cannot always infer the difference of curved lines that are expressed by those equations, although different curves will always yield different equations.[32]

---

[31] As pointed out in Weyl (1927, p. 10) this is evident by a comment Leibniz makes a few pages before the citation given in the text (see Leibniz 1955, p. 70).

[32] "Quae cum infinitis modis variari queant, etiam pro eadem linea Curva innumerabiles aequationes exhiberi poterunt, hancque ob causam ex aequationum diversitate non semper ad diversitatem linearum curvarum, quae illis aequationibus exprimantur concludere licet, etiamsi diversae Curvae perpetuo diversas praebeant aequationes." (Euler 1748, p. 12; my translation; Euler 1990, p. 18, wrongly translates this passage).

Euler then observes that by varying the axis as well as the origin one can express the nature of the same curve in an innumerable number of ways and this leads him to study, although not with these words, an equivalence relation between equations. The ambiguity of the text rests however in the following situation. According to one reading the geometrical curve is primary and thus two equations will be equivalent if and only if they represent the same curve. However, while it is obvious that geometrical curves are prior in an absolute sense, the context of the *Introductio* suggests a different, more radical, interpretation. We have to keep in mind that the *Introductio* is an introduction to algebraic analysis and geometrical curves are not among the objects of this science. Thus, it is more appropriate to read Euler as introducing an 'analytic' object to replace the geometrical object 'curve'. This new object is introduced as the abstract of an equivalence relation between equations where the criterion for two equations to be equivalent is given in formal terms, namely one equation results from the other by an appropriate change of coordinates. This new abstract is nothing else than an algebraic curve (what Euler calls the most general equation) corresponding to the class of equivalence of equations obtainable from one another through appropriate transformations of the coordinates. Thus, although only implicitly, Euler's approach can be characterized as the definition of abstract terms, algebraic curves, given by the following abstraction principle:

$$C(Eq_1) = C(Eq_2) \text{ iff } Eq_1 \sim Eq_2.$$

Let us summarize the situation so far. After having used, in section 1.2, Kronecker as an entry point on the technique of definition by abstraction, I have remarked on the peculiar nature of the 'abstracta' one encounters in number theory in that the functional abstraction outputs an element of the original domain. In section 1.3, I have referred to various cases of implicit use of definition by abstraction in mathematical practice up to Euler and this led to two important distinctions, namely that between thin and thick abstractions and that between relational and substantival readings of definitions by abstraction. We now move to the nineteenth century.

1.4 Definitions by abstraction in number theory, number systems, geometry, and set theory during the XIXth century

When looking for definitions by abstraction in the nineteenth century we are faced with an embarrassment of riches and thus my strategy will be to follow five different areas of research that correspond to major fields of mathematics: a) number theory, b) number systems, c) geometry, and d) set theory. In addition, I will make some brief remarks on definitions by abstraction in physics. Obviously, there are more areas one could look at, e.g. abstract algebra. Needless to say, completeness is not my aim.

We will see that number theory, despite its extensive practice of functional abstraction, is not such a fertile ground for 'thick' definitions by abstraction. But this is not a negative result. On the contrary, it will show 1) why Frege and the Peano school do not cite number-theoretic definitions as cases of definition by abstraction and mention mostly only geometrical and set-theoretic cases and 2) that Frege had access to a long-standing practice of turning equivalence relations into functions whose values, expressed functionally, could be explicitly defined.

1.4.1 Number theory

We have seen that it took a long time to explicitly articulate the notion of 'equivalence relation' as one satisfying reflexivity, symmetry and transitivity. But despite the earlier lack of terminology to single out 'equivalence relations', the classification of objects into classes according to appropriate (equivalence) relations defined on them goes back at least to the late eighteenth century. Consider Gauss's *Disquisitiones arithmeticae* (1801). At the beginning of the *Disquisitiones*, Gauss introduces congruent numbers and moduli. Given integers a, b, c, if a divides $b - c$, then b and c are said to be congruent and a is called their module. Gauss immediately observes that since 0 is divisible by any number, it follows that every number is congruent to itself (with respect to any module whatsoever). By using the sign $\approx$ to denote congruency, Gauss wants to highlight the important analogy between equality and congruency. He does not explicitly mention symmetry and transitivity of congruency but he mentions the following property from which they are derivable (using reflexivity of the congruence relation): if several numbers are congruent with respect to the same one relative to the same module, then they are congruent among themselves (relative to the same module).

The contemporary mathematician would now immediately pass to the equivalence classes, namely the congruence classes modulo n, considered as mathematical objects in their own right. Gauss did not do this, although he of course mentions that the equivalence relation partitions the integers into classes, but he did use equivalence classes as objects in another part of the *Disquisitiones*, namely the part dealing with his theory of composition of forms. Gauss calls a binary quadratic form[33] any equation of the form $ax^2 + 2bxy + cy^2$ where a, b, and c are

---

[33] Throughout this chapter 'quadratic form' will be understood as 'binary quadratic form'.

integers and x, y take values in the integers. He uses as short-hand for a form $ax^2 + 2bxy + cy^2$ the expression (a, b, c).[34] Gauss defines the determinant of the form (a, b, c) to be $b^2 - ac$.[35] The theory assumes that the determinant of the forms under consideration are not perfect squares, for in that case much simpler considerations would apply. In section 222 Gauss refers to Lagrange as having been the first to have carried out general investigations on the equivalence of forms:

In particular he [Lagrange] showed that for a given determinant we can find a finite number of forms so arranged that each form of that determinant is equivalent to one of these, and thus that all forms of a given determinant can be distributed into classes. (Gauss 1801, p. 212 of the English translation)

Here then, attributed to Lagrange, is the notion of class yielded by a specific equivalence. However, one looks in vain in Lagrange for such a notion, which he at best only foreshadowed. However, Lagrange did have the idea of equivalence of forms and for two quadratic forms f and g defined the notion of "forms which can be transformed into one another". It was Gauss who dubbed them 'equivalent forms'. A very clear description of the basic concepts can be found in Dirichlet's lectures on number theory, edited by Dedekind, and in a different context in lectures delivered by Kronecker in 1891 from which I have already quoted. The basic concepts, in all the texts just mentioned, are as follows. Two quadratic forms (a, b, c) and (a', b', c') with the same determinant are equivalent if and only if there exist integers $\alpha, \beta, \gamma, \delta$ such that $\alpha\delta - \beta\gamma = \pm 1$ and

$$a' = a\alpha^2 + 2b\alpha\gamma + c\gamma^2,$$
$$b' = a\alpha\beta + b(\alpha\delta + \gamma\beta) + c\gamma\delta,$$
and
$$c' = a\beta^2 + 2b\beta\delta + c\delta^2.$$

These condition express that we have a transformation[36] of $ax^2 + 2bxy + cy^2$ into $a'x'^2 + 2b'x'y' + c'y'^2$ given by the substitutions $x = \alpha x' + \beta y'$ and $y = \gamma x' + \delta y'$, where $\alpha, \beta, \gamma, \delta$ are integers and with $\alpha\delta - \beta\gamma = \pm 1$.

If the condition is restricted to $\alpha\delta - \beta\gamma = 1$ we have what is called a proper equivalence. It can be shown that in both cases, i.e. $\alpha\delta - \beta\gamma$ equal to ± 1 or equal to 1, the conditions induce equivalence relations on the quadratic forms (with same determinant). Let us consider proper equivalence (i.e. $\alpha\delta - \beta\gamma = 1$).

---

[34] Using an even coefficient $2b$ is something that not everyone in number theory has insisted upon but there are some advantages to having that convention in place.

[35] Some authors call the determinant of the form the discriminant.

[36] In section 54, Dirichlet points out the connection to the theory of transformation of equations under change of coordinates, which we have described when speaking about Euler.

Modifying Lagrange's condition on reduced forms, which guaranteed existence but not uniqueness, it can be shown that for any given binary quadratic form (d, e, f) of determinant $D \neq 0$ there is a unique form (a, b, c) properly equivalent to (d, e, f) satisfying certain arithmetical inequalities on a, b, c. This canonical representative can be computed by an arithmetical procedure, akin to the long division algorithm (see Goldman 2002, 12.4). Consequently, we can effectively test whether two quadratic forms are equivalent. For any determinant D one can thus select what Dirichlet calls a complete system of non-equivalent forms. Thus any form of determinant D is equivalent to one and only one of the forms contained in the complete system of non-equivalent forms. One of the principal results of the theory of quadratic forms is that for any given determinant $D \neq 0$, the number of non-equivalent classes is always finite.

Consider for instance all binary quadratic forms of determinant -3. It turns out that there are only two equivalence classes under proper equivalence and the respective canonical representatives of the two classes are $(1, 0, 3)$ and $(2, 1, 2)$. To test whether two forms whose determinant is -3 are properly equivalent one checks (by an arithmetical algorithm that I have not explained) whether they reduce to the same canonical representative.

The above discussion points to a very important issue. The examples from congruence modulo n and quadratic forms[37] show that the object associated with each original object in the domain of the equivalence relation need not be the equivalence class of objects under the equivalence relation. Rather, one chooses a representative object from the class of equivalence. This is typical of number theory in the nineteenth century where one opts to work with representatives of the classes (what Kronecker called characteristic invariants) as opposed to working with the classes themselves.

Gauss works with representatives both in number congruences (see section 4 of the *Disquisitiones* on least residues) and in the theory of quadratic forms (see section 207). One might think that such a choice is related to ideological qualms about working with infinite classes but that's not the case. Indeed, while Kronecker emphasizes the need to avoid working with infinite classes that's not the case with

---

[37] Kronecker also applied the idea for characterizing the notion of cardinal number as the invariant of the relation of standing in one to one correspondence (he restricts himself to finite sets; no mention is made of infinite sets). He assumes as given the ordinal numbers and his choice of 'reduced' representative for any group [Schar] containing three objects, say, is unsurprisingly given by the collection containing the first three ordinal numbers (his ideas about equivalence and invariance are also contained in his essay on numbers, Kronecker 1887, but not as clearly as they are in the 1891 lectures).

Dedekind, who has taken the 'infinitary turn'[38] and has no such qualms. And yet, in number theory, Dedekind also works with representatives.[39]

By the time Dirichlet gave his lectures on number theory in 1856, edited by Dedekind for the first time in 1863, classes are in evidence both for number congruences and for equivalences of quadratic forms. In the first edition of Dirichlet's lectures we read:

Since any number a is congruent to its remainder r relative to the (positive) modulus k, each number a is congruent to one of the k numbers

$0, 1, 2, \ldots, (k-1)$.

Also, it can be congruent to only one of these numbers, otherwise two of the remainders would be congruent to each other, which is obviously not the case. Therefore, if we divide all the numbers into classes according to the principle that two numbers belong to the same class if and only if they are congruent relative to the modulus k, then the *number* of these classes is obviously k. One of them includes all numbers $\sim 0 (\text{mod } k)$, that is, the numbers divisible by k, the next class contains all numbers $\sim 1 (\text{mod } k)$, and so on.

If one chooses an individual at random out of each class, the resulting system of k numbers has the characteristic property that any integer is congruent to exactly one of these k numbers. [...] All numbers belonging to the same class have many properties in common, so that they behave almost as a single number relative to the modulo k. (Dirichlet 1863, pp. 37–38; 1999, p. 24)

The appeal to choosing specific representatives (at random in the above quotation but the choice here can be made using canonical representatives) leads to the question of whether the classes are intrinsically related to the statement of abstraction principles.

Consider the relation $m \sim n (\text{mod } k)$, i.e. $k/m - n$. It is, as easily shown, an equivalence relation on the integers (zero, positive, and negative). It induces, for

---

[38] One needs to wait until set space the second half of the nineteenth century to encounter a commitment to working with infinite classes with the same legitimacy as finite classes (Bolzano is an earlier exception; and of course there will still be much resistance to this type of infinitistic commitment, for instance by Kronecker). A wonderful passage claiming the right for the mathematician to do so is found in a letter from Dedekind to Lipschitz in June 1876:

> Just as we can conceive of a collection of *infinitely many* functions, which are still dependent on variables, as *one* whole, e.g., when we collect all equivalent forms in a form-class, denote this by a single letter, and submit it to composition, with the same right I can conceive of a system A of infinitely many, completely determined numbers in [the ring of integers] o, which satisfies two extremely simple conditions I. and II., as *one* whole, and name it an ideal. (cited in Ferreirós 2007, p. 110; original in Lipschitz 1986, p. 62).

[39] I should hasten to point out that it is quite clear from the work on ideals that Dedekind is also quite comfortable with treating equivalence classes as objects on which further operations can be defined (see the XIth supplement to Dirichlet's Lectures on the Theory of Numbers).

any choice of k, a partition of all the integers into k classes. For an integer a, the corresponding class is $\mathbf{a} = \{a + nk : n \text{ is an integer }\}$. Of course the curly bracket notation is ours and the real problem here is whether one can interpret the obvious fact about classes of congruences, namely

$$\mathbf{a} = \mathbf{b} \text{ iff } a \sim b(\mathrm{mod}\ n)$$

as the only possible outcome of a definition by abstraction. One is not forced to do so and the situation might more accurately be represented as

$$\alpha = \beta \text{ iff } a \sim b(\mathrm{mod}\ n)$$

where α and β are objects functionally associated to a and b, but not necessarily their equivalence classes. Choosing representatives from the equivalence classes is tantamount to having a function F (which at this stage need not be interpreted set-theoretically) which for each a yields $F(a) = \alpha$, where α is the representative of all objects standing to a in the equivalence relation. In the case of congruence one often chooses the function that yields the least non-negative residues modulo k and thus has range $\{0, \ldots, k - 1\}$.

Is there a substantial difference between the two ways of proceeding? Historically, it is obvious that it was preferable to work with representatives and not with the entire equivalence class as the value of the function. Among other things, a lot of computational information gets lost if one works with the equivalence classes. From a systematic point of view, it seems that in the latter case, even in the absence of an explicit set theory, the totality of the objects introduced with the definition by abstraction can immediately be given an explicit definition—namely as the class of objects standing in the equivalence relation to a given object a—but that in the former case we also have an explicit definition. If the objects associated are as in the cases from number theory we have discussed, then the explicit definition is immediate. In fact the function that takes as input an arbitrary natural number and yields as value the least non-negative residue associated with it, say $F(35) = 0$ when working modulo 5, allows us to explicitly define the value of the function for any input by means of an already known object and thus it amounts to an explicit definition of any expression $F(n)$. When we do not have a definable choice function, we are not anymore in the position to do that.

It is perhaps for the reason just mentioned that we find a split among the early theorizers with respect to definitions by abstraction. While in Weyl (1910) and (1927) number-theoretic definitions count as paradigmatic examples of definitions by abstraction, in the case of Frege and the Peano school the number-theoretic examples are never presented as examples of definitions by abstraction. The motivation for this exclusion, and I will come back to it, would be that a function

$F(x)$ taking arguments in a domain Dom and values also in Dom does not really yield abstracta but rather elements of the same nature we began with. And a function $F(x)$ which takes elements in Dom and yields as values equivalent classes (i.e. special subsets of Dom) might be excluded as a definition by abstraction if one holds, as several members of the Peano school did, that abstracta must be simple.

In conclusion to this section, I can now outline the three theoretical options that are available in the choice of 'abstracta' yielded by a definition by abstraction as far as they have emerged so far.

The first is characteristic of number theory in the nineteenth century and chooses as representative of an equivalence class one of its objects. Usually the choice is specified uniformly by selecting the canonical representative (the least or minimal element in some sense or other).

The second option consists in associating with an element of the domain the equivalence class that contains that element.

The third possibility associates to any object a in the domain an 'abstract object' α, which coincides neither with the class nor with an element of the class.

While the third option is rather common in the nineteenth century (and sometimes, as in Dedekind's theory of irrationals, it is held consciously in opposition to the second possibility), the second option will impose itself slowly but not without resistance, and finding cases preceding Frege's identification of the direction of a line a with the class (or, to use Frege's terminology, the 'extension'[40]) of all lines parallel to it has turned out to be challenging (see Chapter 2). Moreover, as we shall see, it is not always a matter of choice whether one settles on one option or the other. For instance, after the discovery of the set-theoretic paradoxes, defining cardinal numbers by means of the first two options became for a while a hopeless enterprise thereby leading Weyl (1910) and Hausdorff (1914) to embrace option three as the only feasible one. I will come back to this below.

1.4.2 Systems of Numbers and abstraction principles

The development of a rigorous approach to the number systems was not a linear one. Rather, it follows almost a reverse order from the natural one in that mathematicians tried first to account for complex numbers, then irrational numbers, and only later were the integers and the rationals constructed from the natural numbers in the way we have become accustomed to. Since we are here mainly concerned with the use of definition by abstraction before Frege, the two major areas of interest are the constructions of the complex numbers and of the

---

[40] I will return at the end of section 1 of Chapter 2 to the issue of whether Frege's extensions can be interpreted as sets.

irrational numbers. The constructions of the rational numbers and of the negative numbers will here be mentioned only briefly as they are the result mostly of work that went on from the mid-1880s till the end of the nineteenth century. For the complex numbers, I will restrict myself to Hamilton's introduction of couples and to Cauchy's theory of equivalence.

Hamilton introduces couples in his 1837 essay "Theory of conjugate functions, or algebraic couples". Since couples could not at the time be defined set-theoretically, one would expect them to be introduced by means of identity conditions (i.e. by abstraction). Hamilton prefaces his theory of couples by an essay on algebra as the science of pure time in which the key ideas for how to deal with couples are presented in terms of pairs of moments. First Hamilton defines the notion of equality of moments, or equivalence of dates (in itself another wonderful example of the many different contexts in which the word 'equivalence' was used):

If we have formed the *thought* of any one moment of time, we may afterwards either *repeat* that thought, or else think of a *different* moment. And if any two spoken or written *names*, such as the letters A and B, be *dates*, or answers to the question *When*, denoting each a known moment of time, they must either be names of one and the *same* known moment, or else of two *different* moments. In each case, we may speak of the *pair of dates* as denoting a *pair of moments*; but in the first case, the two moments are coincident, while in the second case they are distinct from each other. To express concisely the former case of relation, that is, the case of *identity* between the moment named B and the moment named A, or of *equivalence* between the date B and the date A, it is usual to write

$$B = A; (1.)$$

a written sentence or assertion, which is commonly called an *equation*: and to express concisely the latter case of relation, that is, the case of *diversity* between the two moments, or of non-equivalence between the two dates, we may write

$$B \neq A; (2.)$$

annexing, here and afterwards, to these concise written expressions, the side-marks (1.) (2.), &c., merely to facilitate the subsequent reference in this essay to any such assertion or result, whenever such reference may become necessary or convenient. The latter case of relation, namely, the case (2.) of diversity between two moments, or of non-equivalence between two dates, subdivides itself into the two cases of *subsequence* and of *precedence*, according as the moment B is later or earlier than A. (Hamilton 1837, pp. 299–300)

He then went on to define the equality of pairs of moments as follows:

Considering now any two other dates C and D, we perceive that they may and must represent either the same pair of moments as that denoted by the former pair of dates A and B, or else a different pair, according as the two conditions, $C = A$, and $D = B$,

(9.) are, or are not, both satisfied. If the new pair of moments be the same with the old, then the connecting relation of identity or diversity between the moments of the one pair is necessarily the same with the relation which connects in like manner the moments of the other pair, because the pairs themselves are the same. But even if the *pairs* be *different*, the *relations* may still be the same; that is, the moments C and D, even if not both respectively coincident with the moments A and B, may still be related to each other exactly as those moments, (D to C as B to A;) and thus the two pairs A, B and C, D may be *analogous*, even if they be *not coincident* with each other. (Hamilton 1837, p. 300)

Hamilton is here doing two things. First, he considers the conditions of identity for couples and provides essentially a definition by abstraction where $(a, b) = (c, d)$ iff $a = c$ and $b = d$. Then he immediately goes on to define an equivalence relation $\sim$ over couples, which he calls analogy, whereby $(a, b) \sim (c, d)$ iff $c - a = d - b$. This idea is the key to the introduction of vectors in his work on quaternions. Hamilton then generalizes this to triples and n-tuples and comes back to the theory of number-couples in the Preface to the 1857 *Lectures on Quaternions*.

The theory of number-couples gives Hamilton a way to account for complex numbers, conceived as couples (a, b), where to the complex number i corresponds the couple $(0, 1)$. The theory is well known and I will not say more about it except to add that even if Frege had not read Hamilton, he would have been acquainted with such definitions through Hankel's *Vorlesungen über die Complexen Zahlen und ihre Functionen* (1867). And while this is not the place to write a history of such techniques, I will mention that Heine in 1872 and Weierstrass, in lectures delivered in Berlin in the 1860s, used similar techniques. This type of approach through couples becomes dominant in the foundation of the number systems starting in the mid-1880s, for instance with Stolz's *Vorlesungen* (1885), Jules Tannery's *Leçons d'arithmétique* (1894), and most fully worked out by Couturat in *De l'Infini Mathématique* (1896), Stolz and Gmeiner (1901), and Russell (1903).

Another theory of imaginaries is given by Cauchy who, in 1847, developed a theory of complex numbers by working with polynomials in R[X] modulo $X^2 + 1$, but only using representatives without operations on equivalence classes. Weyl considered this as a paradigmatic case of definition by abstraction and it will be profitable to look at the way he describes it. After having introduced the general notion of equivalence relation, Weyl says:

Now, everytime such a relation between objects $a, b, \ldots$ is defined, one considers it possible to assign to each object a another thing α in such a way that two objects are assigned the same thing if and only if those two objects are equivalent to one another in the sense of the relation $\sim$. Examples in mathematics, where such procedures are used to define a new operational domain of things α, are found galore. I only mention the way in which Cauchy introduces the complex numbers by saying: we consider two polynomials of the independent variable i as equal just in case they are congruent modulo $i^2 + 1$ (i.e. just in case their difference is divisible by the polynomial $i^2 + 1$). In this way emerges the concept of complex number.[41]

Weyl is referring here to the theory of complex numbers presented by Cauchy in 1847 in volume 4 of *Exercises d'Analyse*, namely in "Mémoire sur la théorie des équivalences algébriques substituée a la théorie des imaginaires" (pp. 87–110).[42] He rightly does not claim that Cauchy works with equivalence classes and explains:

It is more satisfactory when the thing α associated to the object a is defined univocally in such a way that the desired invariance comes to light. In the case just mentioned with the complex numbers this can happen easily by associating to each polynomial in the indeterminate i that linear function of i determined univocally, that is congruent to it modulo $i^2 + 1$.[43]

Weyl pointed out that it is sometimes more difficult to find a representative from the class of equivalent objects and mentioned the case of quadratic forms in number theory:

In the arithmetical theory of quadratic forms it is already more difficult to choose, from each class of equivalent forms, a unique (a so-called reduced) form characterized univocally through inequalities.[44]

What would be the reason not to go immediately to the equivalence classes? Here is Weyl's concluding passage concerning this matter in the 1910 article:

Many times one has resorted to defining the associated α as the totality of the objects equivalent to the object a. As long as this concept of totality is logically admissible there is nothing to object despite the psychological monstrosity which is involved in saying: in order to know that these ||| are three strokes I must imagine the totality of all sets that can

---

[41] "Allemal nun, wenn eine solche Beziehung zwischen Objekten a, b, . . . erklärt ist, hält man es für möglich, jedem der Objekte a derart ein anderes Ding α zuzuweisen, daß zwei Objekten dann und nur dann dasselbe Ding zugewiesen erscheint, falls jene beiden Objekte im Sinne der Beziehung $\sim$ einander äquivalent sind. Beispiele in der Mathematik, wo ein solches Verfahren zur Definition eines neuen Operationsbereiches von Dingen α benutzt wird, ließen sich in Hülle und Fülle beibringen. Ich erwähne etwa nur die Art, wie Cauchy die komplexen Zahlen einführt, indem er sagt: Betrachten wir zwei Polynome der unabhängigen Variablen i dann und nur dann als gleich, falls sie modulo $i^2 + 1$ kongruent sind, (d.h. falls ihre Differenz durch das Polynom $i^2 + 1$ teilbar ist), so entsteht der Begriff der komplexen Zahl." Weyl (1910, p. 303).

[42] A clear exposition of the method is given in Grunert's 1857 article "Theorie der Aequivalenzen".

[43] "Befriedigender ist es, wenn in solchen Fällen das dem Objekt a zugewiesene Ding α in eindeutiger Weise so erklärt wird, daß die gewünschte Invarianz zum Vorschein kommt. In dem eben beführten Fall der komplexen Zahlen kann dies einfach so geschehen, daß man jedem Polynom der Umbestimmten i diejenige eindeutige bestimmte lineare Funktion von i zuweist, die ihm modulo $i^2 + 1$ kongruent ist." (Weyl 1910, p. 303).

[44] "Schwieriger ist es bereits in der arithmetischen Theorie der quadratischen Formen, aus jeder Klasse äquivalenter Formen eine einzige durch Ungleichungen eindeutig charakterisierte Form, eine sog. reduzierte, auszuwählen." (Weyl 1910, p. 303).

be correlated in a one to one way with the set of these strokes. However, today it does not appear anymore admissible to construct a set from all the sets that are equinumerous to a given one. One must therefore pursue new ways in order to make the concept of cardinal number a fully determined one.[45]

Weyl's presentation reflects well the state of things at his time. Rather than working with representatives it is also possible to work with equivalence classes, if the concept of totality is logically admissible. And of course we know that in the case of the definition of cardinal number this notion of totality had given rise to trouble and thus defining a cardinal number as the 'set' of all equinumerous sets was no longer an available option (see later section 1.4.4 on set theory). But notice that the objection about using the notion of totality of all three-membered sets does not seem to extend to the use of equivalence classes in other contexts where the threat of paradoxes does not arise (such as the case of quadratic forms in number theory where the equivalence classes are sets of triples of natural numbers). In other words, by 1910, it seems to have been acceptable in the majority of cases (i.e. barring set-theoretic paradoxes) to think of the equivalence classes (as opposed to selected representatives), where appropriate, as the objects yielded by abstraction on an equivalence relation. Of course, Weyl's psychological objection still holds but the acceptance of the logical admissibility of assigning equivalence classes as values of the abstraction function was the consequence of a long development to which I will return in section 1.4.4 and in Chapter 2.

Let us now look at theories of irrational numbers before moving on to geometry and set theory. The importance of real analysis (irrational numbers) and set theory to Frege's understanding of definition by abstraction cannot be exaggerated. It is with reference to Cantor 1883 that Frege introduces the possibility of defining the notion of number through a definition by abstraction. Of course, Frege has no sympathy for the 'psychological' process of abstraction described by Cantor whereby one starts from a given set and then abstracts both from the specific features of the objects and from the order of the elements (see the draft of Frege's critical review of Cantor in Frege 1976, pp. 76–80). But he sees in Cantor's procedure also a logical process, namely the determination of identity conditions

---

[45] "Vielfach hat man auch seine Zuflucht dazu genommen, das zuzuweisende Ding α als die Gesamtheit der mit dem Objekte a äquivalente zu erklären; und so lange dieser Begriff einer Gesamtheit als logisch zulässig gilt, konnte man hiergegen nichts einwenden, trotz der psychologischen Ungeheuerlichkeit, die es etwa involviert, wenn ich sagen würde: Um zu erkennen, daß dieses ||| drei Striche sind, muß ich mir die Gesamtheit aller Mengen vorstellen, welche sich umkehrbar eindeutig auf die Menge dieser drei Striche abbilden lassen. Heutzutage scheint es uns jedoch nicht mehr als statthaft, aus allen Mengen, welche einer gegebenen gleichzahlig sind, selbst wieder eine Menge aufzubauen. Man muß daher andere Wege einschlagen, um den Begriff der Kardinalzahl zu einen völlig bestimmten zu machen." (Weyl 1910, p. 303).

for 'powers' (cardinal numbers) by means of the relation of 'Äquivalenz'. Similar definitional processes had also emerged in the early 1870s in the definition of real numbers and Cantor's theory of irrationals provides a perfect example.

Cantor's theory of irrational numbers was published in 1872, at the same time as other theories developed by Dedekind, Heine, Meray, and others. The paper was titled "Über die Ausdehnung eines Satzes aus der Theorie der trigonometrischen Reihen".

Cantor begins with the domain of rational numbers, which he denotes by A. Then Cantor's proposal can be paraphrased by saying that he extends the concept of quantity to include sequences of rationals

(1) $a_1, a_2, \ldots, a_n, \ldots$ with the following property: for any positive rational ε, there is a natural number n_1 such that $| a_{n+m} - a_n | < \varepsilon$, when $n \geq n_1$ and m is an arbitrary natural number.

He expresses this property of the sequence (1) by the words: "(1) has a definite limit b". The introduction of the limit b must be seen as obtained through an abstraction process. Cantor says:

"These words have at first no other meaning than that of an expression for the property of the sequence and from the circumstance that we associate with the sequence (1) a special sign b it follows that in case of different sequences of such type one must form different signs b, b', b'' etc."[46]

Given two sequences $a_1, a_2, \ldots, a_n, \ldots$ and $a'_1, a'_2, \ldots, a'_n, \ldots$ satisfying the property expressed in 1, and thus with limits b and b', Cantor argues that only three cases, which exclude one another, are possible. Paraphrasing Cantor:

Either $a_n - a'_n$ becomes infinitely small as n increases, or there is a natural number k such that for all $n \geq k$, $a_n - a'_n$ remains greater than a certain positive rational number ε; or there is a natural number k such that for all $n \geq k$, $a_n - a'_n$ remains smaller than a certain negative rational number $-\varepsilon$.

Cantor writes $b = b'$ when the first relation holds and $b > b'$ and $b < b'$ in the other two cases, respectively.

By comparing relations between such a sequence with limit b (what we call a Cauchy sequence) and an arbitrary rational a, Cantor shows that a similar property of trichotomy holds so that either $b = a$ or $b > a$ or $b < a$. Cantor claimed

---

[46] "Es haben also diese Worte zunächst keinen anderen Sinn als den eines Ausdruckes für jene Beschaffenheit der Reihe, und aus dem Umstande, daß wir mit der Reihe (1) ein besonderes Zeichen b verbinden, folgt, daß bei verschiedenen derartigen Reihen auch verschiedene Zeichen $b, b', b'', \ldots$ zu bilden sind." (Cantor 1932, p. 93).

that from the above it follows that the expression "limit of the sequence (1)" has a definite justification although it remained ambiguous whether in addition to claiming the consistency of the introduction of the term "the limit of sequence (1)" within the mathematical language he also was thinking of introducing a denotation for the term.

It is not my intention to enter the question of the shortcomings of Cantor's procedure, which were later addressed also by Cantor himself and have been discussed extensively in the literature on irrational numbers.[47] Rather, I would like to emphasize how the process is intimately tied to a definition by abstraction. One starts from the domain of sequences satisfying the condition expressed in (1). Then one identifies two such sequences $a_1, a_2, \ldots, a_n, \ldots$ and $a'_1, a'_2, \ldots, a'_n, \ldots$ whenever $a_n - a'_n$ tends towards 0 as n increases. If we represent the sequences by s and s' we can write (not Cantor's terminology) $s \sim s'$ whenever the described relation holds. This relation is an equivalence relation. One can now introduce b and b' as abstracta from the sequences and obtain:

$$b = b' \text{ iff } s \sim s'$$

More explicitly one can think of b and b' as being the result of an abstraction given by an operation lim so that $\lim(s) = \lim(s')$ iff $s \sim s'$.

It is tantalizing that Cantor proceeds by saying that "(1) has a definite limit b" is given a meaning through property (1) and thus that by appealing to the equivalence relation expressed by (1) one justifies the propriety of speaking of "the limit of the sequence (1)". The process is very reminiscent of the introduction of singular terms by means of a definition by abstraction about which it is claimed that the meaning of the left-hand side is epistemologically dependent on the right-hand side.

From a Fregean point of view, this is rather unsatisfactory. The limit seems to be introduced only formally, namely as a sign, and this connects Cantor's project to the very similar definition of number given by Heine in 1872 who claims to adopt a "formal standpoint". Heine thought it a virtue of his approach that questions of existence were resolved by appeal to "certain graspable signs" (p. 173). In his case, the functional abstraction that from a sequence [Zahlenreihe] yields its symbol is given by a square bracket notation, namely to a sequence $a, b, c, \ldots$ one associates the sign '$[a, b, c, \ldots]$'. One then provides conditions of identity for sequence-signs in terms of the original sequences.[48]

[47] For an exposition of the alternative theories of irrationals in the nineteenth century see Boniface (2002).

[48] On the similarities between Heine's formal standpoint and Weber's and Hausdorff's formal standpoint, see section 1.4.4.

It also seems to me plausible to interpret Dedekind's theory of irrationals as given by a definition by abstraction.[49] The reason is the following. Recall that Dedekind introduces the irrationals by a process that involves the construction of what we call Dedekind cuts. Dedekind cuts are set-theoretic objects that contain rational numbers as elements. But for reasons having to do with purity of methods (see Reck 2003), Dedekind does not identify the irrationals with the Dedekind cuts. Rather, for every single cut not generated by a rational number he postulates an irrational number, a free creation of the human mind. I claim that this postulation also has the form of a definition by abstraction. Given cuts not produced by rational numbers, C_1 and C_2, we postulate the existence of irrationals associated with these cuts using as condition:

$$\text{irr}(C_1) = \text{irr}(C_2) \text{ iff } C_1 \text{ and } C_2 \text{ are equivalent cuts.}$$

We know nothing more about these irrationals than that they are postulated to be objects whose identity conditions are given by the right-hand side of the above equivalence.[50]

In all these cases, we are within the description of definition by abstraction given by Weyl.

1.4.3 Complex numbers and geometrical calculus

The attempt to develop a satisfactory theory of complex numbers and their geometrical representation led to the notion of oriented segment, 'a new species of magnitude' (Argand 1806, p. 11). An oriented segment has a quantity ('its length') but also a direction. Argand for instance defines the new entities as follows:

In consequence of such considerations it will be possible to generalize the meaning of expressions of the form *AE*, *CD*, *KP*,…and every such expression will designate, in the following, a line of a certain length, parallel to a certain direction, taken in one of the two determinate opposite senses presented by this direction and whose origin is at an arbitrary point while these lines can themselves be the expression of magnitudes of another type. Since they will be the topic of the investigations to follow, it will be appropriate to use a special terminology. We will call them lines in direction [*lignes en direction*] or, more simply, directed lines [*lignes dirigées*].[51]

---

[49] Shapiro, following a suggestion by Burgess, does this in 'Frege Meets Dedekind: A Neologicist Treatment of Real Analysis' (Shapiro 2000).

[50] My point is merely logical and it is consistent with emphasizing the difference between the more ordinary appeal to abstraction principles and the psychological appeal to 'creation' in Dedekind. For an interesting take on this opposition in a discussion of abstraction and structuralism see Linnebo and Pettigrew (2014).

[51] "En consequence de ces réflexions, on pourra généraliser le sens des expressions de la forme *AE*, *CD*, *KP*, … et tout expression pareille désignera, par la suite, une ligne d'une certaine longueur, parallèle à une certaine direction, prise dans un sens déterminé entre les deux sens opposés que

The footnote is tantalizing as Argand says that the expression 'ligne en direction' is short-hand for 'lines considered as belonging to a certain direction'. But the latter expression is not defined. Despite the fact that Argand is aware of introducing a new type of magnitude, he leaves the conditions of identity between directed segments as informally understood and does not provide a formal statement.

I will not here attempt a survey of the voluminous literature concerning the geometrical foundations of complex numbers but rather restrict myself to some major authors such as Bellavitis, Hamilton, and Grassmann.[52] These are authors mentioned and/or discussed in Hankel's *Vorlesungen*, a text Frege knew quite well. In addition, Frege was fully familiar with Grassmann's works, of which more anon.

One would expect to find a good number of definitions by abstraction in the area of the geometrical calculus and vector theory, on account of the need to work with new types of entities such as directed segments etc. But this expectation in some cases is disappointed, for one does not go beyond the equivalence relation which could give rise to an introduction of new entities by abstraction. Consider for instance Bellavitis's 1835 definition of equipollence:

Two straight lines are said to be equipollent if they are equal, parallel, and directed in the same sense.[53]

One does not have here the usual introduction of abstracta. Bellavitis stops at the definition of the equivalence relation of equipollence between the lines and no abstract definition of vector follows on its heels (although one could argue that this is present in practice).

Hamilton also developed his theory of quaternions, presented for the first time in 1844, as a consequence of his interest in the geometrical representation of complex numbers. His approach to quaternions can be described as a generalization of the conception of directed segment which had been employed by several mathematicians in the first three decades of the nineteenth century. Here is Hamilton's definition of vector given in 1866.[54]

---

présente cette direction, et dont l'origine est à un point quelconque, ces lignes pouvant elles-mêmes être l'expression de grandeurs d'une autre espèce. Comme celles doivent être le sujet des recherches qui vont suivre, il est à propos de leur appliquer une dénomination particuliere. On les appellera *lignes en direction* ou, plus simplement, *lignes dirigées*." (Argand 1806, p. 11).

[52] See Freguglia 1992 and 2006 for a clear and accessible introduction to the geometrical calculus developed by these authors.

[53] "Due rette diconsi equipollenti se sono eguali parallele e dirette per lo stesso verso." (Bellavitis 1835, p. 246).

[54] See also Hamilton, *Lectures on Quaternions*, 1853, p. 100 on equality of vectors and pp. 212–213 for equality of arcs in connection to versors.

Section I. On the Conception of a Vector; and on Equality of Vectors. Art. 1. A right line *AB*, considered as having not only *length*, but also *direction*, is said to be a vector. Its initial point *A* is said to be its *origin*; and its final point *B* is said to be its *term*. A vector *AB* is conceived to be (or to construct) the *difference* of its two extreme points; or, more fully, to be the result of the *subtraction* of its own origin from its own term; and, in conformity with this *conception*, it is also denoted by the *symbol B − A*: a notation which will be found to be extensively useful, on account of the analogies which it serves to express between geometrical and algebraical operations. 2. Two vectors are said to be *equal* to each other, or the *equation AB = CD*, or *B − A = D − C*, is said to hold good, when (and only when) the origin and term of the one can be brought to *coincide* respectively with the corresponding points of the other, by *transports* (or by *translations*) *without rotation*. (Hamilton 1866, pp. 1–2)

Hamilton then goes on to prove that vectors satisfy the key identity properties, namely if *CD* and *EF* are vectors which are equal to a third vector then they are also equal to each other. Hamilton's definition is an obvious definition by abstraction. The vector is a function of two points *A* and *B* and the order in which the points are taken is relevant. We can see the definition of vector then as originating from an equivalence relation between segments according to which *AB* is equivalent to *CD* iff they satisfy the condition, let us call it RV(*AB, CD*), given in Hamilton's definition. From there we can go on to abstract Vect(*AB*) = Vect(*CD*) iff RV(*AB, CD*).

Hamilton also provided a definition by abstraction of vector arcs as geometrical representations of versors of quaternions.

Even if Frege had not read Hamilton, he would have learned Hamilton's notions in a book he had read carefully and discussed in the *Grundlagen*, namely Hankel's *Vorlesungen* (1867). Hankel's definition of (directed) segments [*Strecke*] considered as having quantity and direction (p. 74) is given as follows:

One obtains another most important system of operations in geometry when one takes into account at the same time quantity and direction and defines accordingly two *directed segments AB, EF*, in space in general, to be equal if and only if they have the same length, are parallel and have the same direction, that is when *ABFE* form a parallelogram.[55]

On p. 157 the definition of versor is also given by a definition by abstraction.

Let *A, B, ...* be determinate points (different from their opposite points) of a sphere. Then with the *difference* (*B − A*) we will understand the arc *AB* of a main circle lying between *A* and B—the *versor AB*—according to its quantity and insofar as it lies on this main circle. We will set AB = CD when the arc CD is as large as AB and the former directed in the same

---

[55] "Ein anderes für die Geometrie äusserst wichtiges System von Operationen erhält man, wenn man gleichzeitig Grösse und Richtung verwerthet, und dem zufolge zwei *Strecken AB, EF*, allgemein im Raume, dann und nur dann als gleich betrachtet, wenn sie gleich lang, parallel und gleichgerichtet sind, also *ABFE* ein Parallelogram bilden." (Hankel 1867, p. 74).

way as the latter; by contrast, two arcs that do not satisfy the two conditions simultaneously are to be considered unequal.[56]

Also in this case, an abstraction principle is in the offing as we are introducing a new concept, versor, as a function of two points A and B, so that Vers(AB) = Vers(CD) iff the arc(CD) is as large as arc(AB) and AB and CD lie on the same great circle and have the same direction.[57]

I will also recall that Weierstrass, in lectures delivered in 1878, mentions the equality of 'Strecke' in the context of a discussion of one–one correlation for the assignment of cardinal numbers. He says:

Equality. We now call two things a and b equal whenever they are connected, or stand in a relation, let it be expressed by $a = b$, so that $b = a$ also holds and if $a = b$ and $b = c$ then $a = c$ holds too. (For instance, one could call equal two segments in space if they are parallel and have the same sense [nach derselben Seite gerichtet] and can be superposed over each other). Two natural numbers a and b can be called equal when to every unit of a we can correlate a corresponding unit of b, to a different unit of a yet another from b and so forth so that no unit of a is left unmatched in the correlation. (This definition agrees obviously with the general definition of equality given above).[58]

The importance of this passage is related both to the fact that the notion of *Gleichheit* plays an important role and, secondly, to the fact that the definition by

---

[56] "Sind $A, B, \ldots$ bestimmte (von ihren Gegenpunkten unterschiedene) Punkte einer Kugel, so werden wir unter der *Differenz* $(B - A)$ den zwischen A und B liegenden Bogen AB eines Hauptkreises –den *Versor AB*– verstehen, seiner Grösse nach und insofern er in diesem Hauptkreise liegt; d.h.: *wir werden AB = CD setzen, wenn der Bogen CD ebenso gross als AB ist, und mit ihm gleichgerichtet, in demselben Hauptkreise liegt*, während zwei Bogen, welche diesen beiden Bedingungen nicht gleichzeitig genügen, als ungleich angesehen werden." (Hankel 1867, p. 157).

[57] Incidentally, the literature on mechanics also contains such definitions by abstraction. Let me simply refer to the influential textbook Somov (1878–9). There we find the following definition of vectors by abstraction: "Setzen wir fest, dass jede gerade Strecke, die ihre Länge und Richtung oder nur ihre Länge oder nur ihre Richtung stetig mit der Zeit ändert, *eine geometrische Function der Zeit* heissen soll. Nennen wir ferner zwei geometrische Functionen u and u' geometrisch gleich, wenn sie gleiche Länge und einerlei Richtung haben, d.h. wenn sie in einer Geraden oder einander parallel liegen und gleich und von demselben Sinne sind. Eine derartige Gleichheit wollen wir schreiben $\overline{u} = \overline{u'}$ indem wir über jede Seite der Gleichung einen Horizontalstrich setzen." (Somov 1878–9, p. 25). See also Lüroth (1881).

[58] "*Gleichheit*: Wir nennen nun zwei Dinge a und b einander gleich, wenn unter ihnen eine Verknüpfung, Beziehung, sie sei durch $a = b$ bezeichnet, statt findet, daß auch $b = a$ ist und daß, wenn $a = b$ und $b = c$, auch $a = c$ ist. (Z.B. könnte man zwei Strecken im Raume einander gleich nennen, wenn sie parallel und nach derselben Seite gerichtet und zur Deckung gebracht werden können.) Zwei gewöhnliche Zahlen a und b können wir nun einander gleich nennen, wenn indem wir einer Einheit von a eine Einheit von b zuordnen, einer andern Einheit von a eine andere von b u.s.f., jede Einheit von a eine entsprechende von b findet, also keine Einheit von a bei jener Zuordnung übrig bleibt. (Diese Definition stimmt offenbar mit der obigen allgemeinen Definition der Gleichheit überein)." (Weierstrass's lectures on introductory function theory, dated 1878, in Weierstrass 1988, p. 1).

abstraction of equality of 'Strecke' is seen as a similar definition to that of equality of numbers. In Chapter 4, I will show the importance of Schröder's *Lehrbuch der Arithmetik* (1873) in this connection. We will soon see the central role played by the notion of equality and by Grassman in particular.

While it is quite possible that Frege would have been able to see that the definition of vector or versor were similar to those he considered in the *Grundlagen*, we will see that we do not have to rely on a shaky conjecture in order to show that Frege was perfectly aware of the definitions by abstraction that were common in the tradition related to the geometrical calculus. In particular, he was fully familiar with Hermann Grassmann's works in which these definitions take pride of place. Nor was he the only one to see the importance of what Grassmann had done in this connection in the *Ausdehnungslehre*. Only three years after the publication of *Grundlagen*, and I believe independently of Frege, Hermann Helmholtz focused on the importance of definitions by abstraction, and explicitly referred to Grassmann. Let me start with Helmholtz and then move to Grassmann.

Definitions by abstraction were widely used in physics during this period. Indeed, one of the clearest descriptions of what 'abstraction' does is given by Helmholtz in 1887 in *Numbering and Measuring from the Epistemological Point of View*. It is worthwhile to read the whole passage:

The special relationship which may exist between the attributes of two objects, and to which we give the name 'alikeness' [*Gleichheit*], is characterized by Axiom I as already adduced above: *if two magnitudes are both alike with a third, they are alike amongst themselves.* This implies at the same time that the relationship of alikeness is a mutual one. Because from

$$a = c$$
$$b = c$$

it follows just as well that $a = b$ as that $b = a$.

Alikeness between the likenable attributes of two objects is an exceptional case, and can therefore be indicated by factual observation only in that the two like objects, when meeting or interacting in suitable conditions, allow the observation of a particular outcome which does not as a rule occur between other pairs of similar objects. We wish to term the procedure by which we place the two objects in suitable conditions for observing the said outcome, and ascertaining whether or not it takes place, the method of likening [*Vergleichung*].

If this procedure of likening is to give sure information on alikeness or difference for a specific attribute of the two objects, its outcome must exclusively and solely depend upon the condition that both objects possess the relevant attribute in the specific measure, always presupposing that the likening procedure is properly applied.

It follows, from the axioms adduced above, firstly that *the outcome of this likening must remain unaltered if the two objects are interchanged.* It moreover follows that if the two objects *a* and *b* prove to be alike, and it has been found by previous observation using the same method of likening that *a* is also alike a third object *c*, then the corresponding likening of *b* and *c* must also show these to be alike.

These are requirements which we have to lay down for the relevant method of likening. *Only those kinds of procedure which fulfill the said requirements are capable of demonstrating alikeness.* (translation from Ewald 1996, pp. 741–742)

While silent on the obvious requirement of reflexivity, it is clear that Helmholtz is giving the properties that characterize an equivalence relation. Note however, and I have already mentioned this, that we are still missing a general definition of 'equivalence relation' and that the relevant properties are given following the Euclidean model of common notion 1. However, we do have a specific name for the type of relation in question 'alikeness' [*Gleichheit*], which is simply Ewald's choice for translating what normally is translated as 'equality'. Thus, an alikeness relation is one that satisfies axiom 1 as given above. Let's continue with the citation:

From these presuppositions, it follows that 'like magnitudes can be substituted for each other' in the first place for the outcome on whose observation we rely for ascertaining their alikeness. However, the alikeness of further effects on relationships of the relevant objects may also be connected, by natural laws, with alikeness in the case so far discussed, so that the relevant objects may be interchangeable also in these other respects. We are accustomed to expressing this linguistically as follows: we objectivize, as an attribute of the objects, the capacity they have for bringing about the outcome decisive for the first kind of likening; then we ascribe like magnitude of that attribute to the objects found to be alike, and characterize the further effects in which alikeness is preserved as effects of that attribute, or as empirically dependent upon that attribute alone. The sole meaning of such an assertion is always this: that objects which have proved to be alike for the kind of likening which decides concerning alikeness of this particular attribute are also mutually substitutable, without altering the outcome, in further cases mentioned.

Magnitudes whose alikeness or non-alikeness is to be decided by the same method of likening are termed by us '*alike in kind*'. In our separating, by abstraction, the attributes whose alikeness or non-alikeness is hereby determined from everything whereby the objects otherwise differ, there remains for corresponding attributes of different objects only distinction by magnitude. (translation from Ewald 1996, p. 742)

A footnote importantly refers to H. Grassmann's definition of equality [alikeness; *Gleichheit*] to which I will come back momentarily. The above passage is remarkable both for describing the process as a process of 'abstraction' and secondly for pointing out, following Grassmann's exposition in the *Ausdehnungslehre*, the importance of making sure that consideration of the objects under the relevant likeness must yield the possibility of substituting one for the other in the relevant reasonings.

What follows in Helmholtz's article is a list of examples: weights (the weight of *a* equals the weight of *b* iff when *a* and *b* are placed on the pans of a true balance the balance remains in equilibrium); distance between points and length of a segment (given in terms of congruence); time measurement (using physical processes that repeat themselves in like manner exactly and under like conditions); brightness of two areas (by juxtaposition of their boundaries); pitch of tones (using beats); intensities of electrical currents (using the differential galvanometer). Helmholtz summarized the examples thus:

Thus for the task of ascertaining alikeness in various respects, the most varied physical means have to be sought out. But all of them must fulfill the requirements laid down above, if they are to prove an alikeness. The first axiom—'If two magnitudes are both alike with a third, they are alike amongst themselves'—is thus not a law having objective significance; it only determines which physical relations we are allowed to recognize as alikeness. (translation from Ewald 1996, p. 744)

In these deep considerations by Helmholtz we thus find spelled out, in connection to applications to physics, the conception of abstraction that is tied to an introduction of new abstract entities using an equivalence relation. Whether the new entities are new sorts of things or should be thought as being identifiable with previously given objects (such as real numbers) is left open. Weyl (1927) (section 19 "Das Messen") referred to this article by Helmholtz adding a few other examples.

I will not pursue the widespread use of definition by abstraction in physics, whose occurrence and analysis can be traced at least as far back as Mach's use of them in connection to the analysis of the concept of mass in 1868. However, let me point out that the roots of Helmholtz's interest in this matter have to be found in an empiricist attitude he shared with Mach and Maxwell.[59]

---

[59] This is insightfully pointed out in Enriques 1922a, pp. 145–146; pp. 129–130 of the English translation. Enriques explains the issues involved even more perspicuously in 1922b. Enriques mentioned that apart from an early investigation by Apollonius the axiom that 'things that are equal to a third one are equal among themselves' had not been properly scrutinized. The reason, according to Enriques, is that it appears as an analytic axiom which could thus not be denied without contradiction. In order to show how the discussion in physics had questioned the analyticity of the axiom, Enriques went on to discuss the positivist conception of physical theories. Within this context he discussed, among other examples, the concept of mass in Mach in contraposition to that given by Galileo and Newton. Galileo and Newton considered the mass of a body to be the "quantity of matter" (that is the number of atoms, all qualitatively identical, that make up the body). The equality axiom then follows 'a priori'. But Mach refused to accept such a metaphysical definition of mass and insisted on directly comparing two masses by means of an experience (for instance collision or reciprocal action). Enriques concludes: "Under these circumstances it is not anymore a priori evident that the relation designated as "equality of masses" can in fact be considered as "equality of something" in such a way that the property "masses equal to a third one are equal to each other". Thus, the axiom does no longer appear as a logical principle and takes on the role of a physical property of the experimentally defined relation." And in this case, it might fail to hold. As an example Enriques gave the notion of chemical equilibrium: it might happen that a and b, when put in contact under determinate conditions, do not react and the same for a and c, while b and c might give rise to a reaction. He concluded: "thus, the relation that occurs between chemical substances which do not react upon one another cannot—in

Let us go back to pure geometry by following the lead given us by Helmholtz. At the time of writing the *Grundlagen*, Frege could not of course have seen Helmholtz's later essay but he could certainly have had access to the *Ausdehnungslehre*.[60]

Grassmann's *Ausdehnungslehre* (1844) is a great source of definitions by abstraction in geometry and it is with Grassmann that the process of introducing identities based on a relation of equality [being alike], described above by Helmholtz, is brought out to the fore with full clarity. At the very beginning of the *Ausdehnungslehre*, the section on 'Survey of the general theory of forms' (section 1) is devoted to the concept of equality. In that section, Grassmann says:

Here we first establish the concepts of equality and difference.

Since two equals must appear as different in order to stand out as two, and two differents must appear as different aspects of equals, it seems necessary from a superficial consideration to formulate various relations of equality and difference. Thus for example in comparing two line segments one can assert equality of direction or length, of direction and length, of direction and position, and so forth; and in comparing other things, further relations of equality emerge. But already the fact that these relations change according to the character of the things being compared is proof enough that these relations do not belong to the concept of equality itself, but to the objects to which this concept of equality is applied. Thus, for example, for two equally long displacements we cannot say that they are equal as such, but only that their lengths are equal, and so these lengths themselves stand in the absolute relation of equality. Thus we have preserved the simplicity of the concept of equality and can define it: *Those are equal of which one can always assert the same, or more generally what in any judgement can be substituted one for the other.*

It plainly follows from this that, if two forms are each equal to a third, they are also equal to each other; and that those generated in the same way out of equals are again equal. (Grassmann 1995, p. 33)[61]

There is no question that Helmholtz's presentation owes a great deal, as he acknowledges by referring to Grassmann, to this characterization. We will also see in the second part of Chapter 2 that Giuseppe Peano began his reflection on

---

any way—be considered as an equality of some property or element that they have in common". In his conclusion, however, Enriques overstretched his point by claiming that for the logician interested in the theory of equality and definitions by abstraction, it is important to recall that it originates from a philosophical critique of physical theories. But that seems unwarranted as the theory of equality and definition by abstraction, including the awareness that there is no necessity (and thus no analyticity) for the holding of reflexivity, symmetry and transitivity, for an arbitrary relation, was already present in mathematics and preceded Mach's work. On Helmholtz's reliance on Mach and Maxwell see Darrigol (2003), especially pp. 541–545.

[60] I do not know whether Helmholtz knew of Frege's *Grundlagen* in 1887.

[61] That Grassmannian forms were top candidates for a definition by abstraction was very clear to the Peano School. Burali-Forti and Marcolongo (1909) gave the formal definition by abstraction for equality of forms (see appendix to Chapter 2).

definitions by abstraction by meditating on the *Ausdehnungslehre*. What clearly emerges from Grassmann's presentation is that equality can be predicated in many different ways by focusing on particular properties. Consider direction and length (this is the essence of what was already known as a vector). We can compare line segments with regard to both direction and length. Of course the two segments might be different but they can be identified under the relation of having the same direction and length (which, of course, is an equivalence relation). Thus, what is properly said to be identical are abstracta from the original objects. In the case of length, Grassmann says, two segments might be different even though their lengths might be identical or better, to stress the point, absolutely identical.

A good geometrical exemplification of the equality relations discussed above is given by Grassmann in section 14, where the notion of displacement is introduced. Unfortunately, the context in which this takes place would force me to provide several complementary explanations and definitions and I will only cite the following passage because it connects to the issue of directions and parallel lines:

Before we turn to the conjunction of displacements we illustrate the concepts established in the previous paragraph by an application to geometry. The equality of the method of evolution is represented here by equality of direction; the infinite straight line thus appears as a system of first order, the line element as an elementary extension of first order. What there will be called similar appears here as parallel, and likewise parallelism presents two aspects, as parallel in the same and opposite senses. The name displacement we may retain in the corresponding sense for geometry, and thus by equal displacements understand here such line segments as have equal directions and lengths. (Grassmann 1995, p. 48)

This is tantamount to an introduction of vectors by a definition by abstraction. The reason why I don't want to spend too much time in unpacking the above quote is that there is a different text by Grassmann where many more definitions by abstraction are offered and that does not require much preparatory work to be fully grasped. The text is *Geometrical Analysis* published in 1847.

The context in which Grassmann discusses the matter is that of an analysis of the Leibnizian characteristic with the aim of developing a geometrical calculus.[62] Leibniz had used congruence as the basic notion of his characteristic. Grassmann pointed out some alleged inadequacies of the Leibnizian calculus and blamed them on the fact that congruence did not allow, as equality does, the unlimited intersubstitutability of the elements flanking the congruence relation:

Thus we recognize the inadequacy of this characteristic. Wherein does it lie? Obviously, first of all, in that the relation of congruence is introduced to replace the simple relation

---

[62] Whether Grassmann's reconstruction was faithful to Leibniz's project has been forcefully questioned in Etcheverría (1979) but this issue is not relevant for us.

of equality, which alone should occur in mathematical formulas in order that one can substitute throughout. Even though, for this relation of congruence, the principle that what are congruent to the same third thing are congruent to each other is valid, in a congruence relation one can by no means always replace each expression with its congruent.[63] (Grassmann 1995, p. 320)

He then gave an example showing how from the fact that a proposition holds of a segment ab and ab is congruent to bc, it does not follow that the proposition holds of bc. Then he continued:

The possibility of substitution, which is cut off in the above notation, we must however necessarily retain in any mathematical notation if it is to lead to fruitful results. First of all we must therefore change the notation and introduce the essential relation of equality, in which we just set as simply equal that which we can substitute for the other in each proposition.
 If abc [a triangle] is congruent to def, then we ask: What is equal between the two sets of points? Obviously some geometric function of the three points a, b, c must be set equal to the corresponding function of the three points d, e, f. We will tentatively designate this function with the symbol *figure* or *fig.* and instead of $abc \sim def$ now write

$$\mathit{fig}.(a, b, c) = \mathit{fig}.(d, e, f)$$

and likewise

$$\mathit{fig}.(a, b) = \mathit{fig}.(d, e). \text{ (Grassmann 1995, p. 321)}$$

This already provides us with a beautiful example of definition by abstraction. Grassmann gives no indication of how the abstracta are to be interpreted using other available entities. Grassmann goes on to formulate new principles of abstraction in the attempt to provide solutions to the Leibnizian development he was analyzing:

It is then a question of finding the form of this function, specifically for two points the form of the function fig.(a, b). For the equation

$$\mathit{fig}.(a, b, c) = \mathit{fig}.(d, e, f)$$

is just a unification of the three equations

$$\mathit{fig}.(a, b) = \mathit{fig}.(d, e)$$
$$\mathit{fig}.(b, c) = \mathit{fig}.(e, f)$$
$$\mathit{fig}.(c, a) = \mathit{fig}.(f, d).$$

Only if we succeed in ascertaining the form of this function and the laws that underlie it will we be in a position to substitute completely freely. (Grassmann 1995, pp. 321–322)

---

[63] Following Wright 1983 we could say that identity, unlike other equivalence relations, is maximally congruent.

Now Grassmann proceeded to formulate some new abstraction principles from a number of geometrical equivalence relations:

In order to reach this point we will first extend the Leibnizian idea. Thus for example for two similar figures we can say that their shapes [*Gestalt*] are equal, and thus instead of

$abc \sim def$

write

$form(a, b, c) = form(d, e, f).$

Thus we can express the equality of the area between three points or the volume between any four points, or even the affinity, by special symbols. (Grassmann 1995, p. 322)

Grassmann finally defined for "the most general linear relationship", collinearity, the following principle. For two collinear collections a, b, c, d, e, f, and a', b', c', d', e', f', he wrote:

$collin(a, b, c, d, e, f) = collin(a', b', c', d', e', f').$[64]

Later in the text, Grassmann also introduced an equality between quintuples of collinear points writing $collin(a, b, c, d, e) = collin(a', b', c', d', e')$ but what we have seen is sufficient to show the full blown use of definitions by abstraction in Grassmann.

Unlike the case for number theory, the value of the functions involved in these definitions by abstraction is not given either by a class of equivalent objects or by a representative of it. And while those interpretations are not excluded, Grassmann's presentation suggests that we are truly faced with abstracta that are not reducible to anything else. But whatever interpretation one might give of the abstract terms, there is something essential in Grassmann's presentation that needs to be emphasized. Namely, by presenting abstraction principles in a functional form, Grassmann overcame the ambiguity of previous formulations which fluctuated between possible substantival and relational interpretations (see section 1.3). The use of functional terms flanking the equality on the left-hand side of the abstraction principle points to the introduction of new terms, thereby formulating an unambiguous substantival interpretation, and implicitly raising the issue of what they might possibly denote. I claim that these passages from *Geometrical Analysis* are one of the key sources of *Grundlagen* §64. We will come back to this matter in the first part of Chapter 2.

---

[64] It is this particular relation that reveals, as I will argue in Chapter 2, Frege's debt to Grassmann.

1.4.4 Set Theory

It seems appropriate to complete this overview of abstraction principles in the nineteenth century by going back to Cantor, Frege's starting point for his consideration of abstraction principles. Cantor's definition of cardinal and ordinal numbers is too well known to be rehearsed here in detail. I will thus limit myself to recall the main notions that appear in the Cantorian definitions as presented in 1883 and then to point out the foundational qualms that forced a reluctant acceptance of definitions by abstraction during the 1910s on account of the set-theoretical paradoxes. These qualms could be overcome only with the development of a theory of cardinals and ordinals in which the choice of representatives for these numbers could be effected by means of the axiom of choice which allowed a development of the theory of equivalence with canonical representatives. In this way the abstraction principle for cardinals and ordinals moves from a postulation of abstract objects to a choice of representative sets for cardinals and ordinals.

Cantor developed his theory of cardinal and ordinal numbers in a series of articles starting from the mid 1870s that culminated with his *Beiträge zur Begründung der transfiniten Mengenlehre* which appeared in two articles published in 1895 and 1897, respectively.[65] The abstractive definition of cardinal number (in the double sense of relying on psychological processes of abstraction and of using an abstraction principle for their definition) was however implicitly present in the article "Ein Beitrag zur Mannigfaltigkeitslehre" (1878). The opening lines of that article present in fact a definition of cardinality by abstraction:

If two well-defined manifolds M and N can be correlated to one another element by element (something that if it is possible in one way can also happen in many other ways) then one is allowed to use the following expression, namely that these manifolds have the *same power*, or also that they are *equivalent*.[66]

The definition of 'having the same power' is given as an abbreviation for a more familiar relation in terms of one–one correspondence between the collections of elements. The remaining part of the article leaves no doubt that the definition's intention was to license expressions such as 'the cardinality of M equals the cardinality of N'.

---

[65] A useful article in this connection is Deiser (2010).

[66] "Wenn zwei wohldefinierte Mannigfaltigkeiten M und N sich eindeutig und vollständig, Element für Element, einander zuordnen lassen (was, wenn es auf eine Art möglich ist, immer auch noch auf viele andere Weisen geschehen kann), so möge für das Folgende die Ausdrucksweise gestattet sein, daß diese Mannigfaltigkeiten *gleiche Mächtigkeit* haben, oder auch, daß sie *äquivalent* sind." (Cantor 1932, p. 119)

In 1879, in part I of "Über unendliche lineare Punktmannigfaltigkeiten", Cantor presented the same definition given in 1878:

We now reach a completely different but not less important *classification* for linear point manifolds, namely their *power*. In the essay previously mentioned [. . .] we have in general said of two manifolds M and N that belong to two arithmetical, geometrical or any other two sharply defined conceptual realms that they have the same power when it is possible to correlate one to the other according to determinate laws so that to each element of M belongs an element of N and conversely to each element of N belongs one of M. According to whether two manifolds have the same or different power, they can be assigned to *the same class* or *to different classes*.[67]

The notion of equivalence is introduced in the same article:

If two point sets P and Q have the same *power*, and thus belong to the same class (Art. 1, p. 141), then we call them *equivalent* and express this relation by means of the formula:

$$P \sim Q.$$

If one has $P \sim Q$ and $Q \sim R$, then $P \sim R$ is also always the case.[68]

All of this would have been familiar to Frege who explicitly quotes Cantor (1883). Cantor's better known presentation is that involving the double abstraction process presented in *Beiträge zur Begründung der transfiniten Mengenlehre* which appeared in two articles dated 1895 and 1897, respectively. This is a post-*Grundlagen* essay but decisive for the later considerations on how to define cardinals. Let me briefly quote the key passages. Cantor begins with a definition of set or aggregate:

By an "aggregate" we are to understand any collection into a whole M of definite and separate objects m of our intuition or our thought. These objects are called the "elements" of M.[69]

---

[67] "Wir kommen nun zu einem ganz andern, nicht weniger bedeutungsvollen *Einteilungsgrunde* für lineare Punktmannigfaltigkeiten, nämlich zu ihrer *Mächtigkeit*. In der oben angeführten Abhandlung [. . .] haben wir allgemein von zwei geometrischen, arithmetischen oder irgendeinem andern, scharf ausgebildeten Begriffsgebiete angehörigen Mannigfaltigkeiten M und N gesagt, daß sie *gleiche Mächtigkeit* haben, wenn man imstande ist, sie nach irgendeinem bestimmten Gesetze so einander zuzuordnen, daß zu jedem Elemente von M ein Element von N und auch umgekehrt zu jedem Elemente von N ein Element von M gehört. Je nachdem nun zwei Mannigfaltigkeiten von gleicher oder verschiedener Mächtigkeit sind, können sie *einer* und *derselben Klasse oder verschiedenen Klassen* zugeteilt werden." (Cantor 1932, p. 141).

[68] "Besitzen zwei Punktmengen, P und Q gleiche Mächtigkeit, gehören sie also zu einer Klasse (Art. 1, S. 141), so nennen wir sie äquivalent und drücken diese Beziehung durch die Formel aus: $P \sim Q$. Hat man $P \sim Q$ und $Q \sim R$, so ist auch immer $P \sim R$." (Cantor 1932, p. 146).

[69] "Unter einer "Menge" verstehen wir jede Zusammenfassung M von bestimmten wohlunterschiedenen Objekten m unsrer Anschauung oder unseres Denkens (welche die "Elemente" von M genannt werden) zu einem Ganzen." (Cantor 1932, p. 282, transl. p. 85).

After having provided clarifications and examples he went on to define the notion of 'power':

Every aggregate M has a definite "power," which we will also call its "cardinal number."
We will call by the name "power" or "cardinal number" of M the general concept which, by means of our active faculty of thought, arises from the aggregate M when we make abstraction of the nature of its various elements m and of the order in which they are given.
We denote the result of this double act of abstraction, the cardinal number for power of M, by $\overline{\overline{M}}$.[70]

Through a process of psychological abstraction, Cantor ends up formulating an equivalence which counts as an abstraction principle in our sense, namely $\mathrm{Card}(A) = \mathrm{Card}(B)$ iff there is one to one correspondence between A and B. For cardinality he uses a double bar over the set:

Since every element m, if we abstract from its nature, becomes a "unit", the cardinal number $\overline{\overline{M}}$ is a definite aggregate composed of units, and this number has existence in our mind as an intellectual image or projection of a given aggregate M. *We say that two aggregates M and N are "equivalent," in signs*

(4) $M \sim N$ or $N \sim M$,

if it is possible to put them, by some law, in such a relation to one another that to every element of each one of them corresponds one and only one element of the other.[71]

A quick argument then establishes that $\overline{\overline{M}} = \overline{\overline{N}}$ if and only if $M \sim N$. The problem was to say what kind of object the cardinal number of a set M stood for. The psychological process of abstraction that defined the cardinal as a collection of units could not deliver a mathematically convincing answer and thus the definition was immediately perceived as a (thick) definition by abstraction where

---

[70] "Jeder Menge M kommt eine bestimmte "Mächtigkeit" zu, welche wir auch ihre "Kardinalzahl" nennen. *"Mächtigkeit" oder "Kardinalzahl" von M nennen wir den Allgemeinbegriff, welcher mit Hilfe unseres aktiven Denkvermögens dadurch aus der Menge M hervorgeht, daß von der Beschaffenheit ihrer verschiedenen Elemente m und von der Ordnung ihres Gegebenseins abstrahiert wird. Das Resultat dieses zweifachen Abstraktionsakts, die Kardinalzahl oder Mächtigkeit von M, bezeichnen wir mit $\overline{\overline{M}}$.*" (Cantor 1932, p. 282, transl. p. 85).

[71] "Da aus jedem einzelnen Elemente m, wenn man von seiner Beschaffenheit absieht, eine "Eins" wird, so ist die Kardinalzahl $\overline{\overline{M}}$ selbst eine bestimmte aus lauter Einsen zusammengesetzte Menge, die als intellektuelles Abbild oder Projektion der gegebenen Menge M in unserm Geiste Existenz hat. *Zwei Mengen M und N nennen wir "äquivalent" und bezeichnen dies mit*

(4) $M \sim N$ oder $N \sim M$,

wenn es möglich ist, dieselben gesetzmäßig in eine derartige Beziehung zueinander zu setzen, daß jedem Element der einen von ihnen ein und nur ein Element der andern entspricht." (Cantor 1932, p. 283).

cardinality of a set M is introduced by the definition $\overline{\overline{M}} = \overline{\overline{N}}$ if and only if $M \sim N$.[72]

As is well known, the notions of cardinal and ordinal numbers fell hostage, in this naïve conception of set, to several paradoxes, such as Russell's paradox and Burali-Forti's paradox. This had as a consequence a marked disorientation in the 1910s and 1920s as to how best to introduce the notions of ordinal and cardinal numbers. We have already seen that Weyl in 1910 argues that the strategy that identifies the cardinality of a set with the class of objects equivalent to it cannot be held.[73] I would also like to emphasize that Hausdorff (1914) also ended up defending a line of approach that resembled that of Weyl (1910), namely going back to a form of 'thick' abstraction principle where the cardinals are introduced as abstract objects and not defined explicitly in terms of other sets.

Hausdorff described the approach by means of a definition by abstraction as "the formal standpoint" (there are connections to Heine's 1872 work here[74]). After having described the key concepts of one to one correspondence, Hausdorff said:

Sets of a system that are equivalent to a given set and therefore also among themselves have something common that in the case of finite sets is the number of the elements and that in the general case is called the number or the *cardinal number* or the *power*. Concerning the absolute nature of this newly introduced something one can entertain all kinds of different conceptions. G. *Cantor* defines the power of a set as the general concept that originates through abstraction from the individual properties [note 1 omitted] of its elements. B. *Russell* defines it directly as the totality or class of "all" sets equivalent to that set; we consider this objectionable on account of the vague and antinomical properties of this class.

---

[72] A longer treatment of this matter would also have to include Dedekind's partitioning of the universe into classes of sets according to the notion of similarity. In that context, Dedekind also uses the notion of representative of a class of similar sets (see Dedekind 1888, §26).

[73] The strategy goes of course back to Frege and it is found in Russell (1903) where we read: "This method [...], to define as the number of a class the class of all classes similar to a given class [...], is an irreproachable definition of the number of a class in purely logical terms." (Russell 1903, p. 115) At the time, Russell does not seem to have realized that even the definition of the number 1, given in the above terms, leads to an antinomical class just like the Russell class.

[74] Indeed there are interesting connections also to Cantor, Hankel, Thomae and H. Weber. Weber (1906) introduces the notion of finite cardinal using the formal standpoint, namely through a disguised definition by abstraction. Interesting in Weber's approach is also the explicit rejection, as we find it in Weyl and Hausdorff, to appeal to the 'set' of all sets equivalent to a given one. He writes: "Bei der Schaffung des Zahlbegriffs gehe ich aus von einer beliebigen endlichen Menge A und gebe dieser einen Namen oder ein Attribut α, das ich ihre Zahl nenne, und setze dabei fest, daß jede mir etwa vorkommende, *mit A äquivalente Menge* denselben Namen α haben soll, daß aber dieser Name auch *nur* den mit äquivalenten Mengen zukommen soll. Gehe ich statt von A von einer mit A äquivalenten Menge A' aus, und gebe dieser den Namen α, so erhält A denselben Namen. Diese α heißt dann die *gemeinschaftliche Zahl aller mit A äquivalenten Mengen*. Es ist hiernach die Zahl nicht aufzufassen als die *Gesamtheit* oder die *Menge* aller mit A äquivalenten Menge; denn diese Menge kenne ich nicht. Es ist α die *Idee* der durch A bestimmten *Klasse*, wenn ich irgend zwei äquivalente Mengen als derselben Klasse angehörig bezeichne." (Weber 1906, p. 183).

The situation does not become clearer if we consider analogous examples from other areas of mathematics. For, when we ascribe to congruent pair of points a common "distance", to parallel lines a common "direction", to similar figures a common "form", we can also actually specify these concepts by means of segments, angles or numbers. On the other hand one could of course dispense with the concept of power and restrict everything to the consideration of equivalent sets, whereby the ease of expression would naturally be affected.[75]

And then he went on to describe what the formal standpoint amounts to:

We will simply take the formal standpoint and say: to a system of sets A we associate univocally a system of things a in such a way that to equivalent sets, and only to those, corresponds exactly the same thing, that is from $A \sim B$ always follows $a = b$ and vice versa. These new things or signs[76] we call cardinal numbers or powers. We say: A has the cardinality a, a is the cardinality of A, A has cardinal number a, and also (using a as a numeral) A has a elements.[77]

And in a later edition (1927) he wrote:

We shall therefore say in general that equivalent sets have the same *cardinal number or cardinality* (or *power*). That is, we assign an object a in such a way that equivalent sets, and equivalent sets only, have the same object corresponding to them:

$a = b$ means that $A \sim B$

We call these new objects cardinal numbers or powers and say: A has the cardinality a, a is the cardinality of A, A has cardinal number a, [added in the 1935 edition: "a is the cardinal number of A"], and also (using a linguistically as a number) A has a elements.

---

[75] "Mengen eines Systems, die einer gegebenen Menge und damit auch untereinander äquivalent sind, haben etwas Gemeinsames, das im Falle endlicher Mengen die Anzahl der Elemente ist und das man auch im allgemeinen Falle die Anzahl oder *Kardinalzahl* oder *Mächtigkeit* nennt. Über die absolute Beschaffenheit dieses neu eingeführten Etwas kann man allerhand verschiedene Auffassungen hegen. G. *Cantor* definiert die Mächtigkeit einer Menge als den Allgemeinbegriff, der durch Abstraktion von der individuellen Beschaffenheit [note 1 omitted] ihrer Elemente entsteht. B. *Russell* definiert sie geradezu als die Gesamtheit oder Klasse "aller" mit jener Menge äquivalenten Mengen; dies halten wir bei der uferlosen und antinomischen Beschaffenheit dieser Klasse für bedenklich. Wenn wir analoge Beispiele aus anderen Gebieten der Mathematik herbeiziehen, wird die gegenwärtige Situation nicht klarer; denn wenn wir kongruenten Punktpaaren eine gemeinsame "Entfernung", parallelen Geraden eine gemeinsame "Richtung", ähnlichen Figuren eine gemeinsame "Form" beilegen, so können ja diese Begriffe außerdem wirklich durch Strecken, Winkel oder Zahlen präzisiert werden. Andererseits könnte man den Begriff Mächtigkeit freilich entbehren und alles auf die Betrachtung äquivalenter Mengen beschränken, worunter aber die Bequemlichkeit des Ausdrucks erheblich leiden würde." (Hausdorff 1914, pp. 46–47).

[76] Hausdorff's position originates from lectures on set theory he gave in 1910. The evidence is provided in Hausdorff (2013, p. 458).

[77] "Wir werden uns einfach auf den formalen Standpunkt stellen und sagen: einem System von Mengen A ordnen wir eindeutig ein System von Dingen a zu derart, daß äquivalenten Mengen und nur solchen dasselbe Ding entspricht, d. h. daß aus $A \sim B$ immer $a = b$ folgt und umgekehrt. Diese neuen Dinge oder Zeichen nennen wir Kardinalzahlen oder Mächtigkeiten; wir sagen: A hat die Mächtigkeit a, A ist von der Mächtigkeit a, a ist die Mächtigkeit von A, wohl auch (indem wir a als Zahlwort verwenden) A hat a Elemente." (Hausdorff 1914, p. 47).

This formal explanation says what the cardinal numbers are supposed to do, not what they are. More precise definitions have been attempted, but they are unsatisfactory and unnecessary. Relations between cardinal numbers are merely a more convenient way of expressing relations between sets; we must leave the determination of the "essence" of the cardinal number to philosophy.[78]

But while accepting that the 'essence' of cardinal number had to be left to philosophy, Hausdorff recognized, in a note to the 1914 edition, that there was something unsatisfactory in definitions by abstraction:

p. 46. Concerning this definition of cardinality in Cantor, Beiträge I, p. 481 and B. Russell, The principles of mathematics I (Cambridge 1903), p. 115. Cantor denotes according to that double abstraction the power of M with $\overline{\overline{M}}$, the order type with $\overline{M}$. The so-called definition by abstraction (whereby one partitions the things in "classes" on the basis of a symmetrical and transitive relation and then hypostatizes as an independent object of thought all the individual of a class taken together) is very common in mathematics but it is affected often by a certain unclarity.[79]

This problematic state of things among set-theorists can still be seen in Fraenkel's *Zehn Vorlesungen über Mengentheorie* (1927). At first, Fraenkel takes an almost deflationary view claiming that set theory does not aim at offering a definition of cardinal number[80]:

One should certainly not claim that what we have given is a clean "definition" of the concept of cardinal number. By the way, the same criticism can be raised also against other mathematical (or even logical) abstractions, such as for instance against the procedure

---

[78] "Demgemäß sagen wir allgemein, daß äquivalente Mengen dieselbe Kardinalzahl oder Mächtigkeit haben; d. h. wir ordnen jeder Menge A ein Ding a zu derart, daß äquivalenten Mengen und nur solchen dasselbe Ding entspricht:

$a = b$ soviel wie $A \sim B$.

Diesen neuen Dinge nennen wir Kardinalzahlen oder Mächtigkeiten; wir sagen: A hat die Mächtigkeit a, a ist die Mächtigkeit von A, wohl auch (indem wir a als Zahlwort verwenden) A hat a Elemente. Diese formale Erklärung sagt, was die Kardinalzahlen sollen, nicht was sie sind. Prägnantere Bestimmungen sind versucht worden, aber sie befriedigen nicht und sind auch entbehrlich. Relationen zwischen Kardinalzahlen sind uns nur ein bequemer Ausdruck für Relationen zwischen Mengen: das "Wesen" der Kardinalzahl zu ergründen, müssen wir der Philosophie überlassen." (Hausdorff, 1927, p. 25; 2008, p. 69; transl. 1957, p. 28).

[79] "S. 46. Diese Erklärungen der Mächtigkeit bei Cantor, Beiträge I, S. 481 und B. Russell, The principles of mathematics I (Cambridge 1903), S. 115. Cantor bezeichnet im Sinne jener doppelten Abstraktion die Mächtigkeit von M mit $\overline{\overline{M}}$, den Ordnungstypus mit $\overline{M}$. Die sogenannte Definition durch Abstraktion (wobei man auf Grund einer symmetrischen, transitiven Relation die Dinge in "Klassen" einteilt und das den Individuen einer Klasse Gemeinsame als selbständiges Gedankending hypostasiert) ist in der Mathematik typisch, leidet aber häufig an einer gewissen Unklarheit." (Hausdorff 1914, p. 453).

[80] Kamke (1928) takes cardinal numbers to be arbitrary representatives of a class of mutually equivalent sets.

whereby one defines the common characteristic of all lines parallel to each other as their common "direction". There are also serious objections against the proposal to define the cardinal number directly as the *totality* of all sets equivalent to one another, however successful this proposal might be in many analogous cases. Be that as it may this question need not worry us. We do not want to ascribe to cardinal numbers any mysterious essential characteristic, rather only say that, from a purely mathematical standpoint, two cardinal numbers are "equal" or "different". However, according to our understanding, this has the same meaning as the statements that the respective sets are equivalent or not and in this way we have actually given an unobjectionable definition. The objections that have meanwhile been raised against this procedure [Dubislav, Die Definition, second edition, 1926, p. 33] can be dissolved as soon as one becomes clear about the fact that in general one is not after a definition of the concept "cardinal number".[81] Should one wish to introduce this concept itself in mathematics (for instance the concept of finite number) then one must proceed in a completely different way.[82]

The 'completely different' way to proceed required a theory of equivalence developed by means of axiomatic set theory which, through an essential use of the axiom of choice, permitted the identification of cardinals and ordinals with individual sets. Fraenkel gave an outline of such an approach in lecture 8 of his book (pp. 127–134).[83]

---

[81] Here a footnote reads: "Daß trotzdem die Umschreibung von "zwei Mengen sind äquivalent" durch "ihre Kardinalzahlen sind gleich" überaus nützlich sein kann, zeige folgendes Beispiel[. . .]: die Aussage "die Kardinalzahl der Menge M ist kleiner als die der Menge N" ist infolge geeigneten Gebrauchs der obigen Umschreibung (und der damit bequem einführbaren Beziehung des Kleiner-seins) gleichbedeutend mit dem Satzungeheuer: "jede zu M äquivalente Menge ist äquivalent je einer Teilmenge jeder zu N äquivalenten Menge, während keine zu N äquivalente Menge einer Teilmenge irgendeiner zu M äquivalenten Menge äquivalent ist." (Fraenkel 1927, p. 5).

[82] "Man wird nicht behaupten können, daß vorstehend eine saubere "Definition" des Kardinalzahlbegriffs gegeben worden sei. Dieselbe Kritik läßt sich übrigens auch gegenüber anderen Abstraktionen der Mathematik (oder auch der Logik) üben, so z. B. gegenüber dem Verfahren, das gemeinsame Merkmal aller untereinander parallelen Geraden als ihre gemeinsame "Richtung" zu definieren. Auch der Vorschlag, die Kardinalzahl geradezu als die *Gesamtheit* aller untereinander äquivalenten Mengen zu definieren, hat ernste Bedenken gegen sich, so glücklich diese Idee in vielen analogen Fällen auch sein mag. Wie dem auch sei, vom rein mathematischen Standpunkt aus braucht uns diese Frage keine Kopfschmerzen zu Machen. Wir wollen ja von Kardinalzahlen keine geheimnisvollen Wesensmerkmale feststellen, sondern nur Aussagen, zwei solche seien "gleich" oder "verschieden"; das ist aber nach unserer Übereinkunft gleichbedeutend mit den Behauptungen, die zugehörigen Mengen seien äquivalent oder nicht, und hierfür haben wir in der Tat eine zweifelsfreie Definition angegeben. Die gegen dieses Verfahren zuweilen erhobenen Bedenken (z. B. Dubislav 2, S. 33) lösen sich auf, sobald man sich klarmacht, daß eine Definition des Begriffes "Kardinalzahl" überhaupt nicht beabsichtigt ist. Wünscht man diesen Begriff an sich in die Mathematik einzuführen (wie etwa den Begriff der endlichen Zahl), so muß man ganz anders vorgehen (vgl. den Schluß von Vorl. 7/8)." (Fraenkel 1927, p. 5).

[83] This is of course not the place where to give a full account of how in contemporary axiomatic set theory one defines an unobjectionable notion of cardinal number but a brief summary might be useful (I refer to Felgner 2002 for a more detailed account and to Meschkowski 1973 for a more popular account). It seems that Zermelo was the first, in 1915, to have proposed a successful solution to the problem of defining cardinals as specific sets. His solution remained unpublished but it was

This understanding of the formal standpoint is then important for two reasons. First, it can give occasion to reread many of the formalist theories (say, of negative numbers, complex numbers etc.) as theories based on abstraction (for instance, one might think of Hankel's theories). Secondly, it might shed additional light on Frege's worries about definitions by abstraction.

1.5 Conclusion

This first chapter will now come to a close. I think I have established the following theses:

1) There is a much more widespread use of definitions by abstraction in the mathematical literature before Frege than Frege himself acknowledges.

2) The uses of definition by abstraction vary in many ways, but most evidently in the value that the abstraction functions take. In number theory, representatives coming from the set of the original objects are the values of the functions. In synthetic geometry, where the choice of a representative cannot be defined explicitly (unlike in analytic geometry) one works with abstracta. With Frege we will encounter also higher-order abstraction principles (functions from concepts to objects) which will provide even more options than those we have encountered so far.

3) The use of equivalence classes as the values of abstraction functions is unusual before Frege entertains it and this also leads to the question of whether Frege is one of the earliest thinkers who proposed to use the equivalence classes (in the form of 'extensions') as the values of the abstraction function or whether he could have had access to earlier uses of it (more on this in Chapter 2).

We now move to Chapter 2, where I will offer a historical interpretation of section 64 of *Grundlagen* and of some of the related sections, followed by an analysis of definitions by abstraction in the Peano school and in Russell.

---

mentioned in 1940 in publications by Bernays. The first published solution is that of von Neumann (1923) who defined cardinals as specific ordinals. Both Zermelo and von Neumann use the axiom of choice. An alternative approach, which can be implemented within ZF, requires the axiom of foundation and goes back to a suggestion by Dana Scott in 1955 and uses the notion of rank of a set (which itself depends on the construction of the levels V_α through von Neumann ordinals). Scott's approach allows the development of cardinals as sets in the following way: if $r(x)$ denotes the rank of x then $\text{Card}(x)$ is defined as the collection of y's equivalent to x of minimal rank. One then shows that $\text{Card}(x)$ is included in $V_{r(x)+1}$ and is thus a set.

2

The logical and philosophical reflection on definitions by abstraction: From Frege to the Peano school and Russell

In this chapter we will focus on the foundational reflection on definitions by abstraction. The chapter is divided into two main parts. In the first part, devoted to Frege, we will locate Frege's engagement with definitions by abstraction in a Grassmannian tradition and in a debate on directions which was prominent in nineteenth-century German discussions on elementary geometry. In the second part of the chapter, we will move to the foundational analysis of definitions by abstraction in the Peano school and in Russell. We will see that many of the philosophical debates at the time foreshadowed contemporary debates in the neo-logicist literature. I will conclude the chapter with a shorter section containing some considerations on the difference between the nineteenth-century and early twentieth century understanding of definitions by abstraction and those propounded in the neo-logicist literature.

2.1 Frege's *Grundlagen*, section 64

2.1.1 The Grassmannian influence on Frege: Abstraction principles in geometry

On account of having discussed Grassmann in Chapter 1, let me proceed *à rebours* starting from the last part of *Grundlagen* §64.

Let me recall that after introducing the concept of direction, Frege also mentioned that of orientation of planes. He says:

Similarly, from the parallelism of planes, a concept can be obtained that corresponds to that of direction in the case of lines; I have seen the word "orientation"[1] [*Stellung*] used for this. (Beaney 1997, p. 111)

---

[1] It is actually not clear to me why Austin used "orientation" to translate "Stellung" (Beaney 1997 follows Austin in this terminological choice). Much of the literature on projective geometry around

The sources of this attribution can easily be found in two major authorities in projective geometry. But first I would like to claim that Frege had read the relevant passages on shapes and collinearity in Grassmann's *Geometrical Analysis* that I cited in Chapter 1. Towards the end of section 64 Frege says:

> From geometrical similarity there arises the concept of shape, so that, for example, instead of "The two triangles are similar", one says: "The two triangles have equal shapes [*Gestalt*[2]]" or "The shape of the one triangle is equal with that of the other". So too, from the collinear relationship of geometrical figures, a concept can be obtained for which a name has still to be found. (Beaney 1997, p. 111)

The use of collinearity[3] as an equivalence relation that yields abstracta through a definition by abstraction in the same context as the Grassmannian one cannot be a coincidence. In any case, these two instances are the only appeals to collinearity in the context of abstraction principles up to the time of Frege that I know of. Thus, until another plausible source comes up—but Grassmann is so central to Frege that it would be hard to see how any other source could have a claim to be more likely—I think one should assume that Frege had Grassmann's *Geometrical Analysis* present in mind when he wrote this section of *Grundlagen*.

Incidentally, in 1894, Peano says that the definition by abstraction of *Gestalt* has not proved fruitful and immediately afterwards mentions the possibility of an abstraction based on the notion of projectivity (also mentioned by Grassmann in his *Geometrical Analysis*).

What lends further support to the idea that Frege had read Grassmann are the following facts. His discussion takes place within the context of a reflection on identity in Leibniz and the issue of substitutivity salva veritate. §65 of the *Grundlagen* has much in common with Grassmann's reflection on obtaining from congruence an equality that would allow for full substitutivity.

We also have extensive evidence that Frege had worked on Grassmann from early on, for instance in electromagnetic theory, and Jamie Tappenden has shown the important Grassmannian influence that Ernst Abbe, Frege's cherished professor, transmitted to Frege (see Tappenden 2011). Tappenden also pointed out to me that in 1905, just after Abbe's death, Abbe's former research assistant Siegfried Czapski sent around a questionnaire to several people who knew Abbe personally

---

the turn of the twentieth century translates "Stellung" as "aspect". See for instance Reye (1898) and Scott (1900). I will use "aspect" when translating von Staudt and "orientation" in the main text.

   [2] One encounters discussions of "gleiche Gestalt" in other mathematical textbooks such as J. H. van Swinden's *Elemente der Geometrie* (1834). However, the problem is always that it is not clear whether the expression is used substantively as opposed to relationally (namely, "the shape of *a* is equal to the shape of *b* iff . . ." as opposed to "*a* and *b* have the same shape iff . . .").

   [3] The importance of collinearity for Frege's appreciation of the distinction between projective and metrical geometry cannot be overestimated; see for instance his review Frege (1877).

and professionally, asking questions about which of his teachers were the most profound influences, what works influenced him, etc. One of the questions concerned which works most profoundly influenced him (explicitly cast broadly so that it could be scientific, philosophical, social-political, etc.). Frege's answer to that question was just two words: "Grassmanns Ausdehnungslehre" (see Flitner and Wittig 2000, p. 326 for the question and p. 332 for Frege's answer).[4]

Of course, Frege in *Grundlagen* explicitly quotes Grassmann's *Lehrbuch der Arithmetik* (1861) but, to my knowledge, no one had pointed out the dependence of this section of the *Grundlagen* on Grassmann's *Geometrical Analysis*, which Frege never quotes explicitly.

Let us now move to the terminology of direction and orientation. The classical source is von Staudt (1847, p. 17).

40. Parallel planes have something in common[5] which can be grasped in each one of them and can be called its aspect [*Stellung*] so that the aspect of a plane is determined by any plane parallel to it and two planes have the same or different aspects according to whether they are parallel or they intersect each other.[6]

---

[4] Frege also had carefully read texts, such as Hankel (1867), that contain an extensive treatment of Grassmann's key ideas in the *Ausdehnungslehre*.

[5] The expressions "etwas Gemeinschaftliches", which we have already encountered in Kronecker, recalls Frege's discussion of directions and points at infinity in the *Inauguralschrift* of 1873 where he says: "Taken literally, a 'point at infinity' is even a contradiction in terms; for the point itself would be the end point of a distance which had no end. The expression is therefore an improper one, and it designates the fact that parallel lines behave projectively like straight lines passing through the same point. 'Point at infinity' is therefore only another expression for what is common to all parallels, which is what we commonly call 'direction.'" (Frege 1984, p. 1) For a study of the *Inauguralschrift* see Belna (2002). In the *Grundlagen* talk of "points at infinity" is eschewed and I think for reasons that have to do with Frege's unwillingness to accept piecewise definitions. See below my discussion of Frege's correspondence with Pasch. My interpretation thus differs from the six considerations offered in Belna (2002, pp. 401–402).

[6] "40. Parallele Ebenen haben etwas Gemeinschaftliches, was an jeder derselben aufgefasst werden kann und ihre Stellung heissen soll, so dass also die Stellung einer Ebene durch jede zu ihr parallele Ebene bestimmt ist, und zwei Ebenen einerlei oder verschiedene Stellungen haben, je nachdem sie parallel sind oder sich schneiden." (von Staudt 1847, p. 16) Parallels between von Staudt's terminology and Frege's terminology were already pointed out in Belna (2002, pp. 403–404). However, the parallels do not rule out the possibility that Frege's knowledge of von Staudt's work might have been mediated by other standard texts in projective geometry, such as Reye (1866). Wilson (1992) contains an extended discussion of von Staudt's influence on Frege. Wilson sees von Staudt as the source of Frege's grasp of abstraction principles. While it is obvious that von Staudt has something in the vicinity of definitions by abstraction, it is not so obvious to me that von Staudt is such a privileged source for Frege's understanding of such definitions. Rather, Grassmann seems to me a much more likely source. Moreover, the parallel between abstractions for directions and numbers, as we have seen, was already around in quite explicit form before Frege considered it in *Grundlagen*. As pointed out in Chapter 1, Weierstrass, in lectures delivered in 1878, mentions the equality of 'Strecke' in the context of a discussion of one–one correlation for the assignment of cardinal numbers (notice also that the context was that of a definition of *Gleichheit*, just as in Grassmann).

Another possible source is the influential Reye (1866) (*Geometrie der Lage*), where both *Richtung*[7] and *Stellung* [*aspect/orientation*] are discussed.

As it is said of parallel lines that they have the same *direction*, so we say of parallel planes that they have the same *aspect* [*Stellung*]; just as, then, in every direction there lies an infinitely distant point, so for every aspect there is an infinitely distant straight line. All parallel planes in space pass through one and the same infinitely distant straight line, namely, through that straight line in which one of these planes is intersected by each of the others. Parallel planes may therefore be considered as forming a sheaf of planes [*Ebenenbüschel*] whose axis is an infinitely distant straight line; this we shall call a 'sheaf of parallel planes' [*Parallel-Ebenenbüschel*].[8]

Note that in both cases we do not have an explicit definition by abstraction although we are not very far from it. The recognition of something in common to parallel planes leads to the ascription of 'same orientation'. But what is lacking for a full definition by abstraction is the explicit treatment of orientation as a functional term from which to obtain, for planes P and Q, $\text{Or}(P) = \text{Or}(Q)$ iff P and Q are parallel. In fact, the above quotes can still be read relationally, simply as relating P and Q by a predicate that says "P and Q have the same orientation". On the other hand, the introduction of points at infinity and lines at infinity in this connection quickly opens the way for thinking of them as denotata of functional expressions derived from the relations of parallelism for lines and planes, respectively.

Stellung is also used in Baltzer's *Die Elemente der Mathematik*, a well known text at the time, and Baltzer credits von Staudt with the introduction of the terminology.[9]

However, a detailed analysis of Wilson (1992) would lead us too far. I do hope that adding Grassmann to the equation will give us a more balanced appreciation of Frege's sources.

[7] Here is a relevant passage from von Staudt on *Richtung*: "Von den beiden einander entgegengesetzten Richtungen, welche in einer Geraden enthalten sind, wird haüfig nur wie von einer Richtung (mit einem Doppelsinne) gesprochen, so dass alsdann zwei Gerade einerlei oder verschiedene Richtungen haben, je nachdem sie parallel oder nicht parallel sind, und alle nur denkbaren Richtungen durch einen Strahlenbündel, alle Richtungen aber, welche in einer und derselben Ebene sich vorfinden, durch irgend einen in ihr liegenden Strahlenbüschel dargestellt werden." (von Staudt 1847, p. 14).

[8] "Wie von parallelen Geraden gesagt wird, sie haben dieselbe *Richtung*, so sagt man auch wohl von parallelen Ebenen, sie haben dieselbe *Stellung*; gleichwie also in jeder Richtung ein unendlich ferner Punkt liegt, so liegt in jeder Stellung eine unendlich ferne Gerade. Weil sämmtliche parallele Ebenen, die im Raume in irgend einer Stellung denkbar sind, durch eine und dieselbe unendlich ferne Gerade gehen, nämlich durch diejenige, in welcher irgend eine dieser Ebenen von allen übrigen geschnitten wird, so können solche parallele Ebenen auch als ein Ebenenbüschel aufgefasst werden, dessen Axe eine unendlich ferne Gerade ist, und welcher ein Parallel-Ebenenbüschel genannt werden soll." (Reye 1866, vol. 1, p. 17; English transl. Reye 1898, p. 20).

[9] "Nach der Annahme, welche der gemeinen Geometrie zu Grunde liegt (Plani. SS. 2, 8), schließt man: Wenn von zwei Ebenen α, β jede mit zwei nicht parallelen Geraden c, d parallel ist (wenn sie zwei verschiedene Richtungen gemein haben), so sind sie parallel, d. h., sie schneiden sich nicht, sie haben eine *unendlich ferne Gerade* gemein und einerlei *Stellung* [[the footnote refers to von Staudt]].

One ought also to take into consideration that Pasch's *Vorlesungen über neuere Geometrie* (1882) might have been a source for Frege here. However, I am skeptical. There is no evidence that Frege had read this book at the time of working on *Grundlagen* (although he discusses it later in 1904 in correspondence with Pasch; see Frege 1976, pp. 172–174). Moreover, the discussion in terms of *Strahlenbündel* [bundle of rays] and the introduction of points at infinity we find in Pasch does not conform to a definition by abstraction or with the terminology used by Frege in §64. Directions or orientations are not mentioned by Pasch and he, like von Staudt, also does not seem to treat the *Strahlenbüschel* [sheaf of rays] as a new type of entity on which to operate (as a collection of lines considered as one, to use Russell's terminology) but rather distributively (the lines as many, in Russellian terminology). Indeed, in his reply to Frege dated 7 January 1904, Pasch objects to the use of classes in the definition of ideal points, which Frege had apparently suggested, along the lines of his discussion in the *Grundlagen*, in his previous letter. Pasch says that in 1882 he had made use of a definition that used 'a definition of an entire expression [*Redewendung*]'. What is also interesting in this reply by Pasch is that he ascribes to Frege an objection against moving from ordinary points to points at infinity, thereby extending the meaning of 'point' in the process (exactly the process followed by Pasch in his 1882 book). Perhaps for this reason, Frege in §64 does not use the terminology of points at infinity. Frege's reluctance does not seem to have been shared by many other mathematicians.[10] For many contemporaries, parallelism, identity of directions, or intersections at points at infinity boiled down to alternative expressions for the same phenomenon. This is confirmed also in Hilbert's 1891 lectures on projective geometry, where he says that to say that two lines are parallel, or that they have the same direction, or that they intersect at an infinitely distant point are only equivalent ways of speaking (Hilbert 2004, p. 26).

Hätten die Ebenen eine endlich ferne Gerade gemein, so müßte dieselbe sowohl mit *c* als auch mit *d* parallel sein. Also wären *c* und *d* parallel, gegen die Voraussetzung. Bei abstracter Betrachtung, welche von jener Annahme absieht, könnte die gemeinschaftliche Gerade der beiden Ebenen auch in endlicher Ferne liegen, während sie einerseits mit *c*, andrerseits mit *d* parallel ist. In der gemeinen Geometrie wird die Linie der unendlich fernen Puncte einer Ebene als Gerade vorgestellt, weil mit ihr jede Gerade derselben Ebene nur einen (unendlich fernen) Punct gemein hat. Wie von dem unendlich fernen Punct einer Geraden nicht die Entfernung, wohl aber die Richtung, in der er liegt, sich angeben läßt, so kann von der unendlich fernen Geraden einer Ebene nicht die Richtung, sondern nur die Stellung, in der sie sich befindet, angegeben werden. Parallele Ebenen haben einerlei Stellung, gleichwie parallele Gerade einerlei Richtung haben." (Baltzer 1870, p. 145 of volume 2 (third edition, 1870); first edition 1860–1862).

[10] And we have seen that even Frege in the *Inauguralschrift* of 1873 still treated talks of 'direction' and 'point at infinity' as equivalent ways of speaking.

Let me now raise the question of whether Frege's identification of the direction of a line *a* with the extension of the relation 'parallel to line *a*', proposed in §68 of *Grundlagen*, was novel or had any antecedents. The classical sources, such as von Staudt and Reye, cannot be easily interpreted as postulating the existence of the class of all the parallel lines to a given one (and failing this there cannot be an interpretation of direction as the class of all such lines). Rather, the preferred explanation for a *Strahlenbüschel* [sheaf of rays] defines the relation of a line to a *Strahlenbüschel* as 'lying in it' but never seems to make the step of identifying as a new object, called the direction of *a*, the class or the totality of the lines parallel to *a*. One does encounter the *Strahlenbüschel* on which all the lines that have the same direction lie but one does not encounter a direction as an abstract object defined as the collection of the relevant lines. Perhaps this is only conceptual fuzziness on the part of von Staudt but my sense is that the abstract concept of direction defined in terms of a class of lines is not to be found in his work.

But there is a source that Frege had read carefully that comes remarkably close to considering the collection of lines parallel to a given one as a new object to be called 'direction'. I am referring to Eugen Dühring's *Kritische Geschichte der allgemeinen Principien der Mechanik* (1877; first edition 1869), a study of the principles of mechanics. Kreiser (1984) shows that Frege had borrowed this book from the Jena University library and Veraart (1976) lists a five page set of notes that Frege composed on Dühring's book, a set of notes that was unfortunately lost together with other parts of the Frege Nachlaß during the Second World War. Here is the text that impresses me in this connection:

It is not a determinate straight line that forms the essential resting point but rather the general direction which this line represents in space and which is represented by any other line parallel to it as well. If one imagines the totality [*Inbegriff*] of all the possible lines in space that have the same direction then one has a picture for the abstract concept of a direction in space, which is independent of the specific position. The projections for any such direction are always equal no matter which line one chooses as representative of the direction.[11]

It is well known that '*Inbegriff*' in much of nineteenth century mathematics often stands for 'set', 'totality', 'collection', or 'class'. The fact that through the

---

[11] "Nicht eine bestimmte grade Linie bildet hier den wesentlichen Anhaltspunkt, sondern die allgemeine Richtung, welche diese Linie im Raume vertritt, und welche durch jedwede andere ihr parallele Linie ebenso vertreten wird. Man denke sich den Inbegriff aller möglichen den Raum erfüllenden Linien, welche dieselbe Richtung haben, und man hat ein Bild für den abstracten, von der besondern Lage unabhängigen Begriff einer Richtung in Raume. Die Projectionen sind für eine solche Richtung stets gleich, welche Linien man auch zu Repräsentanten der Richtung nehmen mag." (Dühring 1877, p. 274).

representation of all parallel lines (i.e. those having the same direction, according to Dühring) one reaches a concept[12] is not in contradiction with the idea of considering the class of lines (which in Dühring's context are actually segments) with the same direction as a class or completed totality. The latter is said to give a representation (ein Bild) for the abstract concept "of a direction in space". I invite the reader to go back to the definitions by von Staudt and Reye I gave above and the extensive literature on 'Strahlenbüschel' in projective geometry, to verify that this passage by Dühring seems to mark a decisive step in the recognition of direction as represented by the 'Inbegriff' (the set or class), of all lines parallel to a given one. Once again, the reader should not be misled by the fact that the term 'Inbegriff' is also used by von Staudt in giving many of his definitions. An attentive reading of his work shows that directions are never considered as the 'Inbegriff' of the parallel lines.

What motivates my treatment of these matters is the general question, already raised in Chapter 1, of when in nineteenth century mathematics one moves from an equivalence relation to the abstract objects taken as the equivalence classes themselves as opposed to specific representatives, as was the dominant tradition of number theory. Whereas in number theory the choice of a representative can usually be given explicitly, in synthetic geometry (and in the set theory of cardinals and ordinals) there is no way to choose a representative entity except by picking one arbitrarily. In this case the set, or class, or 'extension' in Frege's terminology, seems to be a way to bypass the arbitrariness of the choice. This move to using the 'equivalence class' as the value of the abstraction function seems to have come remarkably late although it is firmly established by the early part of the twentieth century, for instance in the identification of vectors as classes of oriented segments (which we have also found foreshadowed in Dühring). And of course, Russell already in *Principles of Mathematics* (1903) makes extensive use of the technique.

2.1.2 The proper conceptual order and Frege's criticism of the definition of parallels in terms of directions

Almost at the very start of §64, Frege complains about inverting the order between parallelism and direction. He begins by saying:

We split up the content in a different way from the original way and thereby acquire a new concept. Admittedly, the process is often seen in reverse, and parallel lines are frequently defined [by some teachers (*manche Lehrer*), PM] as lines whose directions are equal. The proposition 'If two lines are parallel to a third, then they are parallel to one another' can

---

[12] Frege, in §64, also seems to indicate that one reaches the concept of direction by first having the intuition of parallel lines.

then very easily be proved by appealing to the corresponding propositions concerning equality (of directions). It is only a pity that this stands the true situation on its head![13] (Beaney 1997, p. 111)

The first sentence has been the one that has exercised analytic philosophers. Bob Hale (2001) devoted a whole article (titled "*Grundlagen* §64") to the problem of how to articulate the proper notion of content that could be used to make sense of Frege's claim that the left-hand side and the right-hand side of an abstraction principle share a content that is carved in different ways. I will not enter this difficult,[14] and hitherto not satisfactorily resolved, issue, but simply continue my contextual investigation of §64.

In the above passage, Frege gives us a hint as to what is the background to his statement. 'manche Lehrer' can mean 'some teachers' or 'some instructors'. The translation 'some authorities', which is given by Austin, is completely misleading. Beaney's translation does not fare any better as his translation of the relevant noun phrase eliminates the subject and uses 'are frequently defined' but we are not told by whom. But the explicit mention of 'Lehrer' tells us that the background to Frege's reflection should be found in textbooks related to secondary education. The problem is that the definition of parallels in terms of direction is so widespread in textbooks of geometry used in secondary education in nineteenth century Germany that one feels immediately at a loss. Consider the following examples.

"Two straight lines in a plane which, while not coinciding have the same direction are called *parallel* or *equally directed* [gleichlaufend]"[15]

"Two non-coinciding straight lines[16] in a plane that extended in one and the same direction, or one opposite the other, always keeping entirely one and the same position, no matter how much both sides are extended, are called *parallel*.[17]

---

[13] "Wir zerspalten den Inhalt in anderer als der ursprüngliche Weise und gewinnen dadurch einen neuen Begriff. Oft fasst man freilich die Sache umgekehrt auf, und manche Lehrer definiren: parallele Geraden sind solche von gleicher Richtung. Der Satz: "wenn zwei Geraden einer dritten parallel sind, so sind sie einander parallel" lässt sich dann mit Berufung auf den ähnlich lautenden Gleichheitssatz sehr bequem beweisen. Nur schade, dass der wahre Sachverhalt damit auf den Kopf gestellt wird!" (Frege 1884, §64).

[14] I have found nothing of interest on this matter in the literature preceding Frege.

[15] "Zwei gerade Linien in einer Ebene, welche ohne sich zu decken, gleiche Richtung haben, heissen *parallele* oder *gleichlaufende* Linien" (E. G. Fischer, *Lehrbuch der ebenen Geometrie*, Berlin, 1833, p. 12).

[16] At the time it was not standard to assume the reflexivity of parallelism.

[17] "Zwei in einer Ebene, ohne sich zu decken, nach ein und derselben Richtung hin sich erstreckende oder gegen einander, wie weit man sie auch nach beiden Seiten hin verlängern mag, immer völlig ein und dieselbe Lage habende gerade Linien heissen einander *parallel*" (Grunert, *Lehrbuch der ebenen Geometrie*, Brandenburg a/H. 1870, p. 48).

I can spare myself any more efforts in tracking down more citations because that work has already been done in Heinrich Schotten's book *Inhalt und Methode des planimetrischen Unterrichts. Eine vergleichende Planimetrie* (1890). Schotten discusses the notion of direction in three sections of his two-volume book. First, in volume I (pp. 301–362) and volume II (pp. 3–40), he discusses the matter in connection to the definition of a straight line and in contrast to the notion of 'Abstand' ('distance'); then, he gives an extensive survey of the use of 'direction' in theories of parallels (vol. II, pp. 183–332). The extensive list of definitions of parallel lines in terms of directions can be found in this latter section. These definitions of parallelism are in contrast to those which use the notion of equidistance and which do not rest on the notion of direction. The survey of the sources provided by Schotten, and the accompanying discussion, is also useful as an overview of the foundational debates concerning the relation among the concepts of straight line, direction, and parallel lines. These debates, mostly—but not exclusively, as we shall see—carried out by scholars with an interest in mathematics education, ranged over issues such as whether the notion of a direction of a line is a concept or an intuition, whether it is a priori or a posteriori, whether direction can be conceived without the notion of straight line etc. etc. It is my claim that Frege's comments on direction and parallelism emerge within the background of an extensive discussion over the nature of these basic notions that was quite widespread in nineteenth century Germany.

Having then established the massive presence of the definition of parallel lines in terms of direction in nineteenth century Germany[18] we now have to try to use what Frege says about this matter in §64 to see whether this can lead us to his sources. Let's then consider the theorem that "when two parallel lines are parallel to a third, then they are parallel to each other" and its proof, criticized by Frege, obtained as a consequence of defining parallel lines by means of directions. Incidentally, establishing this theorem would be tantamount to showing that directions satisfy one of the central axioms for being considered magnitudes in their own right. Indeed, one very influential definition of magnitude, given by H. Grassmann in his 1861 textbook of arithmetic (cited by Frege), reads:

Mathematics ($\mu\alpha\theta\eta\mu\alpha\tau\iota\kappa\dot{\eta}$) is the science of the relations between magnitudes. *Magnitude* is said to be anything that can be set equal or unequal to another thing. Two things are *equal* when in any sentence one can substitute the one for the other.[19]

---

[18] French textbooks are markedly different in this respect; on the English side, see Dodgson's [aka Lewis Carroll] *Euclid and its Modern Rivals* (1885, first edition 1879) for an extensive critical examination of the direction theory of parallels.

[19] "Erklärung. Mathematik ($\mu\alpha\theta\eta\mu\alpha\tau\iota\kappa\dot{\eta}$) ist die Wissenschaft von der Verknüpfung der Grössen. *Grösse* heisst jedes Ding, welches einem andern gleich oder ungleich gesetzt werden soll.

This would be satisfied by directions since they can be equal or unequal. Then, the theorem on directions mentioned by Frege would show that directions satisfy the key equality axioms and hence substitutivity salva veritate.

Commentators on §64 have not yet provided examples from the mathematical literature of the period.[20] Let me begin with a textbook not mentioned by Schotten, namely K. E. Zetzsche's *Katechismus der Ebenen und Raümlichen Geometrie* (1878, p. 37). There the theorem has exactly the form criticized by Frege, namely inferring that two parallel lines that are parallel to a third one are parallel among themselves, using the same property for directions.

59. What should be said about straight lines that have a single direction?
   I. If three straight lines G_1, G_2, and G_3 lying on the same plane [. . .] have equal direction then no one of them intersects any of the others.[21]
   II. Two straight lines G_1 and G_2 [. . .] which are parallel to a third straight line G_3 in the same plane, have with G_3 [. . .], and therefore also among themselves, the same direction, and thus are parallel [. . .].[22]

Similarly in 1870 in Zetzsche's *Leitfaden für den Unterricht in der Ebenen und Raümlichen Geometrie*:

Two straight lines which are parallel to a third one in the same plane have equal direction with it and therefore have equal direction also among themselves, and thus are parallel.[23]

But it would be more satisfactory to relate Frege's discussion to authors whose name he mentions and/or whom, on account of their reputation, Frege would have been likely to be acquainted with. Or, even better, it would be good to be able to refer to books we know Frege had used, even though he does not mention them in his correspondence and/or publications. In this particular case, we are in luck.

Gleich heissen zwei Dinge, wenn man in jener Aussage statt des einen das andre setzen kann." (Grassmann 1861, p. 1).

[20] Neither the extended commentary on this section by Thiel in the Centenary Ausgabe of the *Grundlagen* (Frege 1986) nor, just to give only one more example, Kreiser's detailed biography of Frege (Kreiser 2001) provide any such. In addition to the examples mentioned in the text see, among many, also Fischer (1833, section 24, p. 14). Fischer defines parallel lines in terms of direction and then proves the theorem in question using the conceptual dependence criticized by Frege. Killing (1893, pp. 5–6) criticizes the use of direction in theories of parallels.

[21] "59. Was ist über drei Gerade von einerlei Richtung zu sagen?
   I. Haben drei Gerade G_1, G_2, und G_3 (Fig. 34 auf S. 37) in derselben Ebene gleiche Richtung, so schneidet keine die andere." (Zetzsche 1878, p. 37).

[22] "II. Zwei Gerade G_1 und G_2 (Fig. 34) welche einer dritten Geraden G_3 in derselben Ebene parallel sind, haben mit G_3 und daher auch unter sich gleiche Richtung, sind also parallel." (Zetzsche 1878, p. 37).

[23] "Zwei Gerade, welche einer dritten Geraden in derselben Ebene parallel sind, haben mit dieser und daher auch unter sich gleiche Richtung, sind also parallel." (Zetzsche, 1870, p. 13).

In Kreiser (2001, pp. 350–355) we are told that Frege taught in a Privatschule in Jena between 1882 and 1884, that is in the two years leading to the publication of *Grundlagen*. The mathematics textbook adopted in the school was H. Lieber and F. von Lühmann's *Leitfaden der Elementar-Mathematik*, a three-volume work whose first volume is titled *Planimetrie*. The first edition of this book is dated 1876, the fourth edition 1884 and the fifth 1887. Kreiser (2001, p. 354) comments on how Frege would have reacted to the contents of volume 2, which contains the arithmetical material of the course but does not comment on volume 1, which contains the theory of parallels. But it is here that we find the definition of parallels in terms of direction and the statement of the theorem mentioned above just as Frege refers to it in *Grundlagen*. The notion of direction appears from the very start. In §2 (p. 1) the line is defined as the outcome of the motion of a point, a surface through the motion of a line, and a body through the motion of a surface:

A moving point describes a *line*. It has only an extension (length). Should it be limited this occurs through two points. If a line moves in a direction, in which it has no extension, then it describes a *surface*.

[. . .]

If a surface moves in a direction, in which it has no extension, then it describes a *body*.[24]

Then the straight line (*Gerade*) is defined as follows in §3:

If one point moves continuously in the same direction then one calls the line described by it a *straight line* or simply a *line*.[25]

The notion of difference of directions [*Richtungsunterschied*] is used also in defining angles and the concept of direction enters centrally in the proof that all straight [*gestreckten*] angles are equal.

Let us now move to the notion of parallelism. In §14 (pp. 6–7) we have the following definition:

1) Definition. Straight lines that have the same direction are said to be parallel. The sign ||[26] means "is parallel".[27]

---

[24] "Ein sich bewegender Punkt beschreibt eine *Linie*. Sie hat nur eine Ausdehnung (Länge). Soll sie begrenzt werden, so geschieht es durch zwei Punkte. Bewegt sich eine Linie nach einer Richtung, in welcher sie keine Ausdehnung hat, so beschreibt sie eine *Fläche*.[. . .]
Bewegt sich eine Fläche nach einer Richtung, in der sie keine Ausdehnung hat, so beschreibt sie einen Körper." (Lieber and von Lühmann 1876, p. 1)
[25] "Bewegt sich ein Punkt fortwährend nach derselben Richtung, so nennt man die von ihm beschriebene Linie eine *gerade* Linie oder kurzweg eine *Gerade*." (Lieber and von Lühmann 1876, p. 2).
[26] Use of the symbol || for parallel lines is not novel with Frege or this textbook. Its earliest occurrence can be traced back to the Renaissance.
[27] "Erklärung. Gerade Linien, welche gleiche Richtung haben, nennt man parallel. Das Zeichen || bedeutet "ist parallel"." (Lieber and von Lühmann 1876, p. 6).

Then three consequences are stated:

1) Consequence. Parallel lines can never intersect no matter how far they are extended. Were they to intersect then from the intersection point onwards they would have different directions.[28]
2) Consequence. Through a point one can only draw a single parallel to a given line.[29]
3) Consequence. Two lines that are parallel to a third are parallel to each other.[30]

It is obvious that the only argument for proving consequence 3 must use the definition of parallelism in terms of directions and thus Frege's complaint against the reversal of the right conceptual order between parallelism and direction certainly applies here. Indeed, this textbook offered at least two more examples of such inversions, the second of which concerns a notion explicitly mentioned by Frege in section 64, namely that of 'Gestalt'. Consider first how length and congruence are related in the textbook (§28, p. 12):

1) Definition. Two figures are congruent when they can be superimposed on each other, so as to overlap with one another, that is when the boundaries of one with the boundaries of the other completely coincide. One uses the sign ≅ to denote "is congruent".
2) Consequence. Straight lines of equal length are congruent.[31]

While it is true that in this case the notion of length was presupposed from the outset (see the definition of line given above), it is clear that Frege would have found this way of proceeding to sin against the proper conceptual order, for length should be conceptually posterior to the relation of congruence. Consider finally the definition of similarity given in terms of 'same shape' given in §107, p. 54:

Definition. Two polygons are said to be similar when they have the same form [Gestalt], that is when their sides are proportional and their angles are respectively equal. The sign for similarity is ∼.[32]

---

[28] "Folgerung. Parallele Linien können sich nie schneiden, so weit man sie auch verlängern mag. Schnitten sie sich nämlich, so hätten sie von dem Durchschnittspunkte aus verschiedene Richtungen." (Lieber and von Lühmann 1876, p. 7).
[29] "Folgerung. Durch einen Punkt läßt sich zu einer Geraden nur eine einzige Parallele ziehen." (Lieber and von Lühmann 1876, p. 7).
[30] "Folgerung. Zwei Linien, die einer dritten parallel sind, sind einander parallel." (Lieber and von Lühmann 1876, p. 7).
[31] "Erklärung. Zwei Figuren sind congruent, wenn sie sich so auf einander legen lassen, daß sie sich decken, d.h., daß die Begrenzungen der einen mit den Begrenzungen der anderen ganz zusammenfallen. Für "ist congruent" hat man das Zeichen ≅.
Folgerung. Gerade Linien von gleicher Länge sind congruent." (Lieber and von Lühmann 1876, p. 12).
[32] "Erklärung. Zwei Vielecke nennt man ähnlich, wenn sie gleiche Gestalt haben, d.h. wenn ihre Seiten proportionirt und ihre Winkel bezüglich gleich sind. Das Zeichen der Aehnlichkeit ist ∼." (Lieber and von Lühmann 1876, p. 54).

Frege's example concerns triangles instead of polygons but the content is the same. Frege considers that the proper definition of having the same shape is conceptually posterior to the relation of similarity between the figures, exactly the opposite of what one finds in this textbook.

Obviously, this textbook was not the only source of Frege's reflections but it definitely contains many elements that are clearly connected to what Frege says in *Grundlagen* §64 and does have the advantage of being a text Frege would have pondered while teaching in the Privatschule in Jena.

I should add that while all the editions of this textbook that Frege could have used while teaching in the Privatschule and before publishing the *Grundlagen*—namely the first four editions—do not differ on the conceptual order between direction and parallelism, the textbook was revised after the reform of the Prussian teaching plans in 1901. The revision was carried out by Carl Müsebeck in 1902 (Lieber and von Lühmann 1902). In 1902 the definition of parallelism is not given in terms of direction and Müsebeck reorganized the proofs in the book accordingly. But I have no evidence, and I doubt this very much, that this might have been on account of Frege's criticism in *Grundlagen*! It is more likely that the reorganization of the textbook simply reflected Müsebeck's take on the heated debate on the role of 'directions' in the theory of parallels. I now turn to that debate.

2.1.3 Aprioricity claims for the concept of direction: Schlömilch's Geometrie des Maasses

Having shown the existence of a likely source for Frege's criticism of the use of direction in proving theorems about parallels, we have remarked on Frege's thesis that parallelism is prior to direction. As I mentioned, this topic was extensively discussed in nineteenth century Germany. Just to give one instance from the early part of the century, let me mention Andreas Jacobi's theory of parallels (this is not the famous Jacobi but a lesser known mathematician) and the criticism by Sohncke. Jacobi had proposed a theory of parallels that rested on the concept of 'direction' as prior, basic and undefinable.

If someone asks me what direction is and what this terms contains, I will candidly admit that the notion of direction cannot be defined any more than the notion of straight; indeed, the notion of direction and that of straight are identical.[33]

---

[33] The full passage reads as follows: "Si quis ex me quaerat, quid sit directio, et quam notionem haec vox comprehendat, ingenue fateor, notionem directionis non magis, quam notionem recti definiri posse; nam directionis et recti eadem est notio. Ex his statim apparet, unam tantum directionem revera cogitari posse, quia unum genus tantum est linearum rectarum, neque illam, quam antea dixi, incertam vere directionem esse. Inde etiam sequitur, notionem recti non definiri vel explicari posse notione directionis. Ne vero absurdum quid fecisse videar, qui nihilominus explicationem lineae

This theory of parallels, with direction at its core, was criticized by Sohncke who wrote:

This whole matter would be good and useful if one could definitely specify what one must think of the direction of a line. In what Jacobi says the true ground rests on the proposition that one says of lines that they have the same direction when they make equal angles with a third line.[34]

Similar debates took place as a reaction to textbooks by, among others, Ernst Gottfried Fischer (1820_1, 1833_2) and Lorey (1868), to Thibaut's theory of parallels, etc.

The origins of all these developments are to be found in the eighteenth century.[35] At the time very little was known about Leibniz's *analysis situs* (see de Risi 2007) but there were several attempts to demonstrate the parallel postulate through the notion of *situs* (a notion which was known to have been used by Leibniz). Hence the idea came about that parallel lines are those that have the same *situs [Lage]*. Foreshadowed by W. J. C. Karsten (Karsten 1778) the theory became well known through a 1781 essay by Hindenburg (*Ueber die Schwürigkeit bey der Lehre von den Parallellinien*) that was cited extensively and influenced further developments in the late eighteenth and early nineteenth centuries. These developments are at the source of the directional theory of parallels (which is a specific German product). The early essays on the directional theory of parallels, such as Vermehren (1816) or Jacobi (1824), emerge within the context of the previous attempts, such as Hindenburg's, at a theory of parallels with *situs [Lage]* as a central notion.

The key point here is that the notion of identity of situs applied to the theory of parallels led naturally to postulate, or to prove, the transitivity of the relation "to have the same situs". Hindenburg attempts an explicit demonstration whereas others, such as Schwab (1801), take the transitivity for granted. But the transitivity of parallels is equivalent to the parallel postulate (since it is false in hyperbolic

---

rectae a directionis notione petierim, haec monere liceat. Unaquaque enim lineae rectae explicatione animus non demum cognoscit vel cogitat rectam, sed recognoscit tantum, quod jam antea sibi finxit [...] Num vero ad recognoscendam notionem recti aliqua explicatio lineae rectae melior atque aptior fit, quam ea, quae a directionis notione petita est, vix dicere ausim. Huc accedit, quod illa explicatio unica est, qua difficultas tolli possit, quae adhuc in parallelarum theoria reperiebatur. Lineae rectae vero, quae infinite productae sibi non incidunt, appellantur parallelae vel (uti eas nominare ex antecedentibus nobis licet), lineae unius ejusdemque directionis." (Jacobi 1824, p. 37).

[34] "Diese Sache wäre ganz gut und brauchbar, wenn man bestimmt angeben könnte, was man sich unter der Richtung einer Linie zu denken habe. In dem, was Jacobi sagt, liegt offenbar der Satz zum eigentlichen Grunde, daß man von Linien sagt, sie haben gleiche Richtung, wenn sie mit einer dritten Linie gleiche Winkel machen." (Sohncke 1838, p. 382).

[35] For the next two paragraphs on the early history of the directional theory of parallel lines, I am indebted to Vincenzo de Risi whose help is gratefully acknowledged.

geometry). Thus Hindenburg thought he had proved the parallel postulate and Schwab (and others) claimed to have shown its plausibility or even to have provided a "philosophical" or "logical" proof based on considerations related to the relation "identity of situs". The directional theory of parallels is thus mainly old wine in a new bottle: many of the authors who write about the directional theory of parallels simply call direction [*Richtung*] what earlier had been called situs [*Lage*] but they maintain the same aim, namely to prove or make plausible the parallel axiom starting from a new conceptual definition of parallel lines.

These developments were obviously known to the mathematical community at large in particular because the only public statement made by Gauss on non-Euclidean geometries (or better the impossibility of proving the parallel postulate) is a review, published in 1816, of a book by Schwab written in 1814.[36] In that review Gauss complained that the trick of demonstrating the parallel postulate from the transitivity of the relation "have the same situs" was useless and non-geometrical. We will see that, modulo the change from "have the same situs" to "have the same direction" these are criticisms that recur in later literature (see Gugler vs. Schlömilch below) and that debate, directly or indirectly, influenced Frege, or so

---

[36] This review (Gauss 1816) has been partially translated in Ewald (1996). Regrettably, the parts that are our main concern were omitted from the translation. I add the text for the reader's convenience. "Der Verfasser der erstern Schrift hatte bereits vor 15 Jahren in einer kleinen Abhandlung *Tentamen novae parallelarum theoriae notione situs fundatae* einen ähnlichen Versuch gemacht, indem er Alles auf den Begriff von Identität der Lage zu stützen suchte. Er definirt Parallel-Linien als solche gerade Linien, die einerlei Lage haben, und schliesst daraus, dass solche Linien von jeder dritten geraden Linie nothwendig unter gleichen Winkeln geschnitten werden müssen, weil diese Winkel nichts anders seien, als das Maass der Verschiedenheit der Lage dieser dritten Linie von den Lagen der beiden Parallel-Linien. Diese Beweisart ist in der vorliegenden neuen Schrift wiederholt, ohne dass wir sagen könnten, dass sie durch die eingewebten philosophischen Betrachtungen an Stärke gewonnen hätte." Gauss then reproaches Schwab for having defined position [*Lage*] as a relational concept but as having used it as an absolute concept: "Wenn wir von des Verfassers Definition: Situs est modus, quo plura coexistunt vel iuxta se existunt in spatio ausgehen, so ist Lage ein blosser Verhältniss-Begriff, und man kann wohl sagen, dass zwei gerade Linien *A, B* eine gewisse Lage gegen einander haben, die mit der gegenseitigen Lage zweier andern *C, D* einerlei ist. Aber der Verf. gebraucht das Wort Lage in seinem Beweise als absoluten Begriff, indem er von Identität der Lage zweier nicht coincidirenden geraden Linien spricht. Diese Bedeutung ist offenbar so lange leer und ohne Haltung, bis wir wissen, was wir uns bei einer solchen Identität denken und woran wir dieselbe erkennen sollen. Soll sie an der Gleichheit der Winkel mit *einer* dritten Linie erkannt werden, so wissen wir ohne vorangegangenen Beweis noch nicht, ob eben dieselbe Gleichheit auch bei den Winkeln mit einer vierten geraden Linie Statt haben werde: soll die Gleichheit der Winkel mit *jeder* andern geraden Linie das Criterium sein, so wissen wir wiederum nicht, ob gleiche Lage ohne Coincidenz möglich ist. Wir stehen mithin *nach* des Verf. Beweise noch gerade auf demselben Punkte, wo wir *vor* demselben standen." (Gauss 1816, in Gauss 1873, p. 365) The emphasis on identity conditions for the notion of *Lage* points to the literature on parallel lines in the period 1770–1820 as a possible fruitful area of investigation for definitions by abstraction which would give us the bridge between the seventeenth century sources (Desargues, Leibniz) and the mature uses of definitions by abstraction we found in Grassmann. Moreover, approaches such as Schwab's foreshadow further developments such as that of Schlömilch to be analyzed next.

I will claim. We can now move straight to what I consider the most interesting debate on such matters in connection to Frege.

In order to venture a conjecture as to whether Frege is here reacting specifically to some parts of this debate on directions (as opposed to simply reflecting on the specific definition found in a textbook he was familiar with), we will have to connect Frege's comments to authors he was likely to have been influenced by. My tentative conjecture is that there is an author who, more than anyone else I have seen, fits the bill. The author I have in mind is Oscar Xaver Schlömilch (1823–1901).[37]

In section 27 of the *Grundlagen*, Frege quotes Schlömilch's *Handbuch der algebraischen Analysis* (1845). Schlömilch was also the author of a book titled *Grundzüge einer wissenschaftlichen Darstellung der Geometrie des Maasses* (1849, reprinted seven times until 1888) that is relevant for us. Although I have no smoking gun to show a direct acquaintance of Frege with this book, I would like to mention six facts in support of the plausibility of this conjecture.

First, Schlömilch uses the concept of 'direction' as the foundational concept for the theory of parallel lines and develops a calculus of directions. Second, the reactions to Schlömilch's approach (for instance that of Reuschle in *Zeitschrift für das Gesammtschulwesen*; see below) contain many points of contact with what Frege says in *Grundlagen*. Third, Schlömilch had taught in Jena and had been part of a Jena school of mathematics (although he had left in 1849, see Kreiser 2001 pp. 75, 77) and had even corresponded with Frege as editor of the *Zeitschrift für Mathematik und Physik* in 1881. Although Schlömilch had left Jena in 1849, he had been a colleague of K. Snell who was first a teacher and then a colleague of Frege. Fourth, we know that Frege had taken out of the Jena University library, in 1874, yet another book by Schlömilch (*Übungsbuch zum Studium der höheren Analysis*, 1873; see Kreiser, 1984, p. 21). Fifth, Schlömilch's *Grundzüge einer wissenschaftlichen Darstellung der Geometrie des Maasses* was extremely successful and went through seven editions between 1849 and 1888 (see Cantor 1901 for a detailed list of the editions). And, last but not least, in the plethora of publications on elementary geometry up to Frege that use the word "gleiche Richtungen", Schlömilch seems to be the only one who transformed the relation of 'having the same direction' into an equality by writing 'Richtung $(a) =$ Richtung (b)' thereby resolving the ambiguity between the relational and substantival construction of

---

[37] See http://www-history.mcs.st-and.ac.uk/Biographies/Schlomilch.html for details on his central role in German mathematics at the time. On his work see Cantor (1901) and for Schlömilch's philosophical interests and his relation to Fries see Hermann (2000).

'two lines have the same direction'; moreover, he even attempted to use this equality in a deductive argument.

Let us look at Schlömilch's text in more detail. Schlömilch's book is meant as a geometrical handbook with the ambitious aim of contributing to a reform [*Umgestaltung*] in the teaching of geometry. Indeed, the reviewer Reuschle got so annoyed at the claims of reform contained in the preface that in his review he talks about "the reproaches that he [Schlömilch] with the tone of a true Messiah of geometry raises against all geometries that have appeared so far" (Reuschle 1850, p. 248). Among its novelties is the theory of parallel lines. In §1 Schlömilch defines a line by making the notions of 'direction' and 'position' central.[38] He says:

The only feature we grasp in such a straight line is its direction; at the same time the knowledge of the direction is not sufficient to determine with certainty the straight line, for there can obviously be several straight lines that have the same direction without being identical to that one. Indeed, one obtains such straight lines whenever starting from different points of the plane one proceeds in one and the same direction. By contrast, if in addition to the direction of the straight line we also know the point from which it originates or through which it passes then there can be no more doubt about the position of the straight line, i.e. *A straight line is determined in its position [Lage] as soon as a point on it and its direction are given.*[39]

A first consequence is that all lines that have the same direction and that go through the same point coincide. Moving on in §2 to two straight lines, Schlömilch says that considering only the direction there are only two possible cases, namely either both have the same direction or they have different directions. He then

---

[38] The definition given by Schlömilch is also found in Wolff ("A straight line is described by a point that continuously keeps the same direction (*Richtung*), and therefore all points on a straight line lie in the same place (*Gegend*).")

[39] "Das einzige Merkmal, welches wir an einer solchen Geraden wahrnehmen, ist ihre Richtung; gleichwohl aber reicht die Kenntnis dieser Richtung nicht hin, um die Gerade selbst so unzweifelhaft zu bestimmen, dass man sie von jeder anderen Geraden sogleich unterscheiden könnte, denn es kann offenbar mehre Gerade geben, welche dieselbe Richtung besitzen, ohne deshalb mit jener völlig einerlei zu sein, und man erhält in der That solche Gerade, wenn man von verschiedenen Punkten des Raumes aus jedesmal nach einer und derselben Richtung fortgeht. Ist dagegen ausser der Richtung der Geraden noch der Punkt bekannt, von welchem sie aus oder durch welchen sie hindurch geht, so kann kein Zweifel mehr über die Lage der Geraden sein, d. h. *Eine Gerade ist ihrer Lage nach bestimmt, sobald ein Punkt in ihr und ihre Richtung gegeben sind.*" (Schlömilch 1849, p. 8) It is interesting that in Frege's 1873 *Inauguralschrift* we find exactly the same claim: 'As a straight line is determined by two points, it is also given by a point and a direction. This is only an instance of the general law that, whenever we are dealing with projective relationships, a direction can represent a point (Frege 1984, p. 1). However, I do not think that this is in itself evidence for the acquaintance of Frege with Schlömilch's book, for such characterizations were rather widespread in 'directional' accounts of parallels.

defines two lines as parallel when they have the same direction.[40] Relying on the notion of direction he then proved that two parallels never meet. The proposition that for any line and a point outside of it, there is always a parallel to the line passing through the point is obtained by asserting that the parallel line is determined as soon as we have one point (lying outside the given line) and a direction (given by the first line). In the case in which the lines have different directions then they must meet at a point. In a footnote, Schlömilch comments on the priority of directions over parallelism:

In both propositions discussed above:

1) Straight lines with the same direction do not meet,
2) Straight lines with different directions always meet, it is assumed that the equality or difference of directions is known by nature (*a priori*) and that one distinguishes, relying on it, whether the straight lines meet or not.[41]

Here then is a splendid case of the reversal of the conceptual order about which Frege complained. For even though Schlömilch continues the passage saying that it would also be possible to proceed from the non-intersection or the intersection of the straight lines to decree the sameness or difference of directions, he thinks the first approach is the natural one. In other words, Schlömilch has the equivalence corresponding to the abstraction principle on directions but he starts from the equality of directions to obtain the parallelism of the lines. As we shall see, Schlömilch not only speaks of 'Gerade von gleicher Richtung', he will also explicitly write down an equality sign between directions (thereby, through this substantival use of 'direction', turning directions into names of objects).

Next, the notion of lines meeting at a point O is used to explain the notion of an angle as the difference between the directions of two lines. The notion of direction is also used to define that of rotation [*Drehung*] and several properties of angles are exhibited using the notion of rotation. Two angles that originate by equal rotations in the same sense are equal. Then Schlömilch defines addition, subtraction, multiplication and division of angles. The definition of right angle and straight angle is then offered together with some properties of adjacent angles.

---

[40] "Zwei gerade Linien, welche gleiche Richtung besitzen, ohne in einander zu fallen, wie z.B. *AB* und *A'B'* in Fig. 3, heissen *Parallelen*, was durch *AB* || *A'B'* bezeichnet ist." (Schlömilch 1849, p. 10)

[41] "In den beiden oben ausgesprochenen Sätzen:

1) Gerade von gleicher Richtung treffen nicht zusammen,
2) Gerade von verschiedenen Richtungen treffen immer zusammen,

ist angenommen, dass man die Gleichheit oder Ungleichheit der Richtungen von Hause aus (*a priori*) kenne, und man entscheidet daraus das Zusammentreffen oder Nichtzusammentreffen der Geraden." (Schlömilch 1849, pp. 10–11)

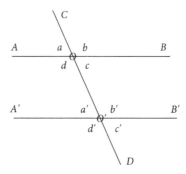

Figure 2.1 Diagram 7 from Schlömilch's *Grundzüge einer wissenschaftlichen Darstellung der Geometrie des Maasses*.

Section 3 is the most interesting for us. Here Schlömilch looks at the relation between three lines on the plane with respect to direction. The possibilities, he says, are the following: either the three lines have one and the same direction, or two have the same direction and one does not, or all three have different directions. He focuses his attention on the second case. This is the case that we normally study when we look at parallel lines that are intersected by a third line. The angles that originate in this configuration have familiar relations of equality:

The corresponding angles are

a and *a'*
b and *b'*
c and *c'*
d and *d'*.

Alternate angles are

a and *c'*
b and *d'*
c and *a'*
d and *b'*.

The first two are interior and the second two external alternate angles. Finally external and internal angles are paired as follows:

a and *d'*
b and *c'*
c and *b'*
d and *a'*.

It is at this point that Schlömilch begins working with the equalities for directions. Indeed under the assumption that AB and $A'B'$ have the same direction he immediately writes

"the direction of $OA =$ the direction of $O'A'$"

and then he continues using such equations to prove that corresponding angles are equal, alternate angles are equal, and a variety of other standard theorems about external and internal angles. Consider the proof for the statement that corresponding angles are equal.

The relations holding between these angles can easily be discovered in the following way. Since according to the assumption AB and $A'B'$ have the same direction, then we have

the direction of $OA =$ the direction of $O'A'$;

and since, moreover, OC and $O'C$ are parts of one and the same line, then it must be the case that

the direction of $OC =$ the direction of $O'C$.

[The first edition continues as follows (the text was removed from further editions): "If we subtract the second equation from the first keeping in mind that equal minuends taken from equal subtrahends yield equal differences, then it follows that the difference between the directions of OA and OC is equal to the difference between the directions of $O'A'$ and $O'C$. By means of the definition of angle this means nothing else than $\angle AOC = \angle A'O'C$ or, more briefly, $a = a'$. (1849, p. 17)]

By means of the proposition that equals compared to equals yield equals also follows that the difference between the directions of OA and OC is equal to the difference between $O'A'$ and $O'C$. The first difference of direction will be indicated through the angle $A'O'C$, and since both differences of direction are equal, then we must have $\angle AOC = \angle A'O'C$ or, more briefly, $a = a'$. In likewise simple manner one can arrive at the equations $b = b', c = c'$ and $d = d'$. Accordingly, one is justified in stating the propositions: *The corresponding angles are equal with one another.*[42]

---

[42] "Die Beziehungen, welche zwischen diesen Winkeln statt finden, lassen sich auf folgendem Wege leicht entdecken. Da nach der Voraussetzung AB und $A'B'$ gleiche Richtung besitzen, so ist auch

die Richtung von $OA =$ der Richtung von $O'A'$;

da ferner OC und $O'C$ Teile einer und derselben Geraden sind, so muss

die Richtung von $OC =$ der Richtung von $O'C$;

sein;

[[the first edition continues as follows (the text was removed from further editions): "mittelst des Grundsatzes, dass Gleiches mit Gleichem verglichen Gleiches liefert, folgt hieraus, dass der Unterschied zwischen den Richtungen von OA und OC gleich ist dem Unterschiede zwischen den Richtungen von $O'A'$ und $O'C$. Vermöge der Definition des Winkels heisst dies aber nichts Anderes als $\angle AOC = \angle A'O'C$ oder kürzer $a = a'$.]]

We see here that not only the equality of directions is considered as an identity, so as to preserve substitution, but that Schlömilch also starts computing with directions as if they were magnitudes that can be added and subtracted.

These early sections of Schlömilch's book were derivative on a previous textbook, K. Snell's *Lehrbuch der Geometrie* (1841). Snell was a professor of Frege in Jena[43] and a teacher and later colleague of Schlömilch; incidentally, the third part of Snell's textbook, *Stereometrie* (1857), was written by another disciple of Snell who went on to be a colleague of Frege for many years, namely Heinrich Schäffer. For instance, Snell also begins by asserting that

The straight line has only two properties, which can be specified more precisely, namely length and direction. We ask through which means both of them, length and direction, are determined.[44]

Yet, despite the centrality of the notion of *Richtung*, Snell does not change the theory of parallels into a theory founded on the notion of *Richtung* but indifferently talks of parallel lines and *gleichlaufend* lines. At times he speaks of lines as having the same direction (see p. 25) but without turning this into an equality the way Schlömilch does. This points thus to a certain originality in Schlömilch's approach.

2.1.4 The debate over Schlömilch's theory of directions

The conceptual ordering between parallelism and direction chosen by Schlömilch was something he was criticized for in a review published in 1850 in the *Zeitschrift für Gesammtschulwesen* by Prof. Reuschle (from Stuttgart) and in an article, published in the same issue containing Reuschle's review, by Prof. Gugler (also

---

ziehen wir die zweite Gleichung von der ersten ab und berücksichtigen dabei den Grundsatz, dass gleiche Minuenden um gleiche Subtrahenden vermindert gleiche Unterschiede übrig lassen, so folgt auf der Stelle, dass der Unterschied unter den Richtungen von OA und OC gleich ist dem Unterschiede unter den Richtungen von $O'A'$ und $O'C$.

Die erste Richtungsdifferenz wird durch den Winkel AOC angezeigt, die zweite durch den Winkel $A'O'C$, und da beide Richtungsunterschiede gleich sind, so muss auch $\angle AOC = \angle A'O'C$ oder kürzer $a = a'$ sein. Ebenso leicht kann man zu den Gleichungen $b = b'$, $c = c'$ und $d = d'$ gelangen; man ist hiernach berechtigt, den Satz auszusprechen: *Die korrespondierenden Winkel sind einander gleich*." (Schlömilch 1849, pp. 16–17; 1888, p. 17).

[43] Frege spoke of Snell as his "revered" teacher. See Tappenden (2011) and Frege (1979, p. 280) and (1984, p. 127). Incidentally, Snell, like Schlömilch, was influenced by Fries' *Naturphilosophie*.

[44] "Die grade Linie hat nur zwei Eigenschaften, welche einer näheren Bestimmung fähig sind, nämlich Länge und Richtung. Wir fragen also, wodurch werden beide, Länge und Richtung, bestimmt." (Snell 1841, p. 17).

from Stuttgart) on the foundations of elementary geometry. The latter asserted among other things, the unviability of using direction as the primitive notion on which to base a theory of parallels.

Reuschle's review is titled "Über die synthetische Methode in der Mathematik mit Rücksicht auf Schlömilchs *Grundzüge einer wissenschaftlichen Darstellung der Geometrie des Maßes*". The review is almost thirty pages long and covers a lot of ground but I will focus only on what Reuschle says on directions and parallelism. This discussion starts on p. 253 and continues until the end of the review (p. 258). Reuschle begins by presenting the theory, more or less as I have done above, but providing extensive verbatim citations from Schlömilch's text. Reuschle rejects at the outset the definition of straight line given by Schlömilch in terms of direction that I have already cited above. He summarizes Schlömilch's position with the statement that the one feature of a line that we immediately perceive is that of direction. Then he outlines Schlömilch's theory of parallels. The critical remarks begin after the presentation of the proof, given above, that corresponding angles are equal, which appeals to equality of directions but also adds and subtracts directions. Here is Reuschle's scathing appraisal:

One does not believe one's eyes and there is no need to add anything else. In section 6, Schlömilch handles directions as magnitudes that can be added and subtracted; but what would be the unit if one wanted to express these magnitudes through numbers? But that is not enough. The person who is used to seeing a magnitude of the same type originate from addition and subtraction of magnitudes witnesses the following unheard of spectacle, namely that subtraction of directions yields angles. I don't know whether the mathematician should be more astonished about this or the logician about the other side of the matter, i.e. the four-pronged inference—"Herod was a fox"—about the confusion of an arithmetical difference, namely the difference which arises through subtraction, with another difference. Indeed, even if one is entitled to look at the difference of directions as an angle, this is of course no difference in the *magnitude* that can be determined through subtraction.[45]

---

[45] "Man traut seinen Augen kaum, und kaum ist es nöthig noch etwas beizufügen. In Nr. 6 behandelt also Schlömilch Richtungen als Größen, welche addirt und abgezogen werden können; was wäre aber die Einheit, wollte man diese Größen durch benannte Zahlen auszudrücken? Aber nicht genug: für den, welcher gewohnt ist, aus Addition und Subtraktion von Größen stets eine Größe gleicher Art entspringen zu sehen, ereignet sich das unerhörte Schauspiel, daß Subtraktion von Richtungen Winkel gibt, und ich weiß nicht, soll hierüber mehr der Mathematiker erstaunen, oder der Logiker über die andere Seite der Sache, über die Viergliedrigkeit des Schlusses—"Herodes war ein Fuchs"—über die Verwechslung eines arithmetischen Unterschieds, der durch Subtraktion entstehenden Differenz, mit einer sonstigen Verschiedenheit, denn wenn man zugesteht den Winkel als Unterschied von Richtungen anzusehen, so ist dies doch offenbar kein Unterschied in der Größe, welcher durch Subtraktion ermittelt werden könnte." (Reuschle 1850, p. 256).

Even a charitable reading of what Schlömilch is up to, adds Reuschle, would show that he is presupposing the statement that parallel lines (or lines with the same direction) are those that intersected by a third line yield equal angles.

There is also a further major weak point in Schlömilch's approach to parallels, according to Reuschle. It is the demand that 'direction' is given intuitively a priori with full clarity:

This already leads to the other point, which is just as bad, of the theory of parallels put forth by Schlömilch. The author does in no way aim at this explanation [of lines having the same direction], rather he *downright* requires the *intuition of straight lines having the same direction* just as *Euclid* requires that of the [indefinite] extension of a straight line. The concept of direction is in general for him a fundamental one, in itself clear and in need of no further explanation, on which the concept of straight line itself must rest.

Since however he is not completely at ease with the matter, so he feels compelled to the undoubtedly very *pedagogical* note—that both in the definition of parallels as well as in the theorem (above in n. 4) on lines having different directions [...]—*it be assumed that the concept of equal or unequal direction is a priori*! If in n. 6 [the proof of the equality of angles] we have a *non plus* of inferential connectedness, we have in n. 4 [the assumption of a prioricity for the recognition of same direction] the equivalent for conceptual clarity.[46]

A reflection on the foundational situation leads Reuschle to claim, justly and in ways that recall Gauss's criticism of Schwab, that what we have here is the use of hidden premises that when brought to light give us back Euclid's approach:

If one is to think something determinate or clear under the notion of equal direction then this can obviously only be that such equal directions deviate by the same amount from a fixed basic direction, that is to say that parallel lines are those that make equal angles when intersected by a third line. But then *Schlömilch's* basic proposition is not essentially different from the Euclidean, the old disastrous *axiom eleven*. For it says: If two straight lines make unequal angles with a third line then they meet if extended sufficiently (but if the angles [opposite or alternate] are unequal then the internal angles on one side are smaller than two straight angles). However, by means of this assumption, Euclid proves everything rigorously, although according to Schlömilch he is awfully unscientific, in order to prove first that the external angle in the triangle is greater than each internal opposite

---

[46] "Allein dies führt zu dem anderen, fast ebenso schlimmen Punkt in der *Schlömilch*'schen Parallelentheorie. Der Autor hat keineswegs diese Erklärung im Auge, vielmehr fordert er die *Anschauung gleichgerichteter Gerader schlechtweg* wie *Euklid* die Verlängerung einer Geraden; der Begriff der Richtung ist ihm überhaupt ein ursprünglicher, für sich klarer, nicht weiter zu erklärender, auf welchem derjenige der geraden Linie selbst beruhen soll. Weil ihm aber doch nicht ganz wohl bei der Sache ist, so fühlt er sich zu der gewiß äußerst *pädagogischen Note* gedrungen, bei der Erklärung der Parallelen, sowie bei dem Grundsatz (oben in Nr. 4) über ungleich gerichtete Gerade, [auf dessen Aufstellung als des "einzig möglichen" er sich in der Vorrede nicht wenig zu Gute thut,] sei *angenommen, daß der Begriff der gleichen oder ungleichen Richtung ein apriorischer sei*! Haben wir in Nr. 6 ein Non plus von Schlußbündigkeit, so haben wir in Nr. 4 ein Ditto von Begriffsklarheit." (Reuschle 1850, p. 257).

angle and then immediately after that it is equal to the internal angles taken together. By contrast, he (Schlömilch), by means of his rigorous ordering of propositions, could prove the latter statement straight away.[47]

This leads to also claiming the circularity[48] (or simply the identity with the parallel postulate as formulated by Euclid) of the assumptions made in the assertion that given a straight line with given direction and a point outside of it we can give a line of same direction as the given one that passes through that outside point. In conclusion, it is the concept of direction that is absolutely unsuitable to serve as the foundation for a theory of parallel lines and that reveals itself as a *cul de sac*:

Not only is the above mentioned proposition *begged* but all the rest too, for since n.1 [definition of straight line in terms of direction] is begged then this *other proposition* already actually means nothing else than: through a point there is only one parallel to a [given] straight line; and with all the begged propositions nothing comes out because the basic concept, namely the *concept of equal directions*, in Schlömilch's development lacks any clarity. This in the end is also the $\pi\rho\omega\tau\sigma\nu$ $\psi\epsilon\upsilon\delta\sigma\varsigma$, for it is just the concept of equal direction that for the theory of parallels is a *cul de sac*, as it can be seen. To clarify this comprehensively would lengthen this already almost too detailed essay and would be suited, with a dialectical elucidation of the theories of parallels in general, for an independent essay.[49]

We will find also in Gugler the claim that Schlömilch's approach simply begs the question and is founded on a pretense. Remarkably, Frege expresses a very similar

---

[47] "Wenn man unter Geraden von gleicher Richtung etwas Bestimmtes oder Klares denken soll, so kann es offenbar nur das sein, daß solche gleiche Richtungen von einer Grundrichtung um gleichviel abweichen, d.h. daß Parallelen solche Geraden sind, die mit einer dritten schneidenden Geraden gleiche Winkel machen. Dann aber ist *Schlömilchs* Grundsatz von dem euklidischen, dem alten unheilvollen *elften Axiom* nicht wesentlich verschieden; denn er heißt: wenn zwei Gerade mit einer dritten ungleiche Winkel machen, so treffen sie genugsam verlängert zusammen (sind aber die Winkel [Gegen- oder Wechselwinkel] ungleich, so sind die inneren an einer Seite kleiner als zwei Rechte). Mit Hülfe dieser Voraussetzung beweist aber Euklid Alles vollkommen streng, obwohl er nach Schlömilch so graülich unwissenschaftlich ist, um erst zu beweisen, der Außenwinkel im Dreieck sei größer als jeder innere Gegenwinkel, und hintendrein dann, daß er beiden zusammen gleich sei, während er (Schlömilch) mittelst seiner streng systematischen Anordnung gleich das Letzere beweisen konnte." (Reuschle 1850, pp. 257–258).

[48] Note that the line of argument is similar to that used by Sohncke against Jacobi mentioned earlier.

[49] "*Erbettelt* aber ist nicht nur der eben genannte Grundsatz, sondern auch alles übrige, weil Nr. 1 erbettelt ist, denn dieser *andere Grundsatz* heißt eigentlich bereits nichts anderes als: durch einen Punkt gibt es zu einer Geraden nur eine Parallele; und es kommt mit all' den Bettelsätzen erst nichts heraus, weil der Grundbegriff, nämlich der *Begriff der gleichen Richtungen* in der *Schlömilch*schen Aufstellung aller Klarheit ermangelt. Dies ist auch zuletzt das $\pi\rho\omega\tau\sigma\nu$ $\psi\epsilon\upsilon\delta\sigma\varsigma$, denn eben der Begriff der gleichen Richtungen ist für die Parallelentheorie eine Sackgasse, wie man es auch angreifen mag. Dies umfassend zu erörten, würde den bereits fast zu ausführlichen Aufsatz noch länger machen und eignet sich mit einer dialektischen Beleuchtung der "Parallelentheorien" überhaupt zu einer eigenen Abhandlung." (Reuschle 1850, p. 258).

opinion (he speaks of "Erschleichung", i.e. something obtained by false pretense or a fraudulent acquisition) when discussing those positions that use direction to ground parallelicity.

Let me move now to Gugler's essay "Über die Begründung der Elementar-Geometrie". This is also a very long essay but what concerns us is treated in a section titled "Parallel lines are lines which have the same direction (without coinciding)". This section occurs in the early part of the essay where Gugler is showing the defects of many definitions given in geometry. Schlömilch is not named explicitly but that he is the main (or one of the main) intended targets is obvious from the reference to a priori intuition of directions and from the fact that Gugler was certainly au courant of Reuschle's essay (they were both in Stuttgart and writing for the same journal).

Gugler's essay is just as interesting as the former material we have analyzed. The key criticism against Schlömilch's theory of parallels is given as follows:

It rests on an extension obtained through abstraction of the original concept *direction*, which at first was connected to *one* specific line. In order to attain that abstraction the theory of parallels must first be taken care of [*abgethan sein*].[50]

What Gugler is saying is that even granting an intuition of the direction of a single line (which would have to be an individual intuition), the passage to the concept "direction" can only be obtained through an abstraction which requires the theory of parallels. Let's read again Frege's words:

It is only a pity that this stands the true situation on its head! For everything geometrical must surely originate in intuition. I now ask whether anyone has had an intuition of the direction of a line. Of the line, certainly! But is the direction of a line distinguished in intuition from the line itself? Hardly! This concept [of direction] is only found through a mental act that takes off from intuition. On the other hand, one does have an idea of parallel lines. The proof just mentioned only works by covertly presupposing [*durch eine Erschleichung*], in the use of the word 'direction', what is to be proved; for were the proposition 'If two lines are parallel to a third, then they are parallel to one another' false, then $a//b$ could not be transformed into an equation.[51] (Beaney 1997, p. 111)

---

[50] "Er beruht auf einer durch Abstraction gewonnenen Erweiterung des ursprünglichen Begriffs *Richtung*, welcher zunächst an *eine* bestimmte Gerade geknüpft war. Um zu jener Abstraktion zu gelangen, muß zuvor die Parallelentheorie abgethan sein." (Gugler 1850, p. 265).

[51] "Nur schade, dass der wahre Sachverhalt damit auf den Kopf gestellt wird! Denn alles Geometrische muss doch wohl ursprünglich anschaulich sein. Nun frage ich, ob jemand eine Anschauung von der Richtung einer Gerade hat. Von der Gerade wohl! Aber unterscheidet man in der Anschauung von dieser Gerade noch ihre Richtung? Schwerlich! Dieser Begriff wird erst durch eine an die Anschauung anknüpfende geistige Thätigkeit gefunden. Dagegen hat man eine Vorstellung von parallelen Geraden. Jener Beweis kommt nur durch eine Erschleichung zu Stande, indem man durch den Gebrauch des Wortes "Richtung" das zu Beweisende voraussetzt; denn wäre der Satz "wenn zwei

Here Frege's criticism goes beyond Gugler's in that Gugler grants an intuition of direction for a specific line but claims that in order to get to a concept 'direction' or 'equality of direction' one needs the theory of parallels (hence, presumably, the concept of 'parallel' lines). Frege rejects even the intuition of the direction of a single line but rather thinks that one can reach the concept of 'direction' by going first through an intuition, namely the intuition of parallel lines (described as a 'Vorstellung' but in context with the previous line it is clear that this is the intuition necessary for the 'geistige Thätigkeit' to operate).

Gugler then points out that the definition of 'Richtung' seems to solve in one go all the difficulties but that all of this is mere deception.

It cannot be denied that the above explanation is very impressive, for—put together with the right definition of angle,—it seems to remove all difficulties of the theory of parallels in the simplest and most natural way. But it is all appearance.[52]

What is the reason for the deception (recall that Frege speaks of an 'Erschleichung')? Once again we are back to intuitions and concepts:

At the beginning we merely know that two lines have different directions when they originate from a [common] point and do not coincide; and they have *the same* direction if they coincide (and are considered in one and the same sense). For the difference or identity (equality) of directions of two lines originating from different points (before developing the theory of parallels and the proposition on the angles of a triangle that is connected to it) we have no direct criterion. We infer only indirectly a *difference* in directions when we experience that the lines intersect somewhere, from which then follows that their extensions (from the point of intersection onward) start from *one* point without coinciding. However, whether *the equality* of the directions is—directly and *exclusively*,—grounded by the fact that the lines *do not* intersect, remains undecided as long as the proposition that two lines with different directions (in the plane) *must* intersect cannot be proven. And not only is this proof impossible but rather the proposition itself cannot even be stated since we learn about the difference of the directions only through the occurrence of the intersection point and thus the proposition would say nothing more than: lines that (considered as endless) intersect do intersect.[53]

Geraden einer dritten parallel sind, so sind sie einander parallel" unrichtig, so könnte man $a//b$ nicht in eine Gleichung verwandeln." (Frege 1884, §64).

[52] "Es läßt sich nicht laügnen, daß obige Erklärung sehr besticht, da sie, zusammengestellt mit der richtigen Definition des Winkels, alle Schwierigkeiten der Parallelentheorie auf's Einfachste und Natürlichste zu heben scheint. Allein es bleibt beim Scheine." (Gugler 1850, p. 265).

[53] "Zu Anfang wissen wir blos, zwei Gerade haben verschiedene Richtung, wenn sie von einem Punkte ausgehen und nicht zusammenfallen; und sie haben *dieselbe* Richtung, wenn sie aufeinander-fallen (und in einerlei Sinn betrachtet werden). Für die Verschiedenheit oder Einerleiheit (Gleichheit) der Richtungen zweier von verschiedenen Punkten ausgehenden Geraden haben wir (vor der Parallelentheorie und den mit ihr zusammenhängenden Sätzen von den Dreieckswinkeln) gar kein unmittelbares Kriterium; wir schließen erst mittelbar auf eine Richtungs-*Verschiedenheit*, wenn wir erfahren, daß die Linien sich irgendwo schneiden, indem dann ihre weiteren Fortsetzungen (über

Now comes the part that rejects the a priori intuition of direction and that makes it clear that Schlömilch, I claim, is the intended target:

Briefly: Without consideration for the existence of an intersection point, we can *a priori* neither speak of lines of *equal* direction nor of lines of *unequal* direction. Rather we have at first merely lines that intersect and lines that do not intersect. Should one thus want through the previous definitions of parallels to avoid the appeal to the infinite extension of the straight lines [. . .] then this is a vain attempt.[54]

The conclusion goes on to assert yet another circularity in the account:

The second (and most probably the decisive) reason for that definition [in terms of direction] is however the concern for the theory of parallels. If one has inferred from the equality of directions of two lines that the angles that originate through a third line are equal, then one, immediately after, *proves* that those lines do not intersect. But if the direct assumption of equal direction in different lines is already in itself, on account of its meaninglessness, not allowed, then even less it can be allowed to build on it conclusions of such encompassing importance.[55]

I will close here the treatment of the critical reviews of Schlömilch's book. As I said, I have no way to prove that Frege was aware of these sources. However, it is not unlikely that through his conversations with Snell, one of his teachers and later colleague, and/or Schäffer, this debate about directions and parallels might have been brought to his attention. But at the least, I feel confident in asserting that Frege's reflections on direction and parallelism emerged within the context of a lively discussion in Germany over such matters.

---

den Schnitt hinaus) von *einem* Punkte ausgehen ohne zusammenzufallen. Ob aber die *Gleichheit* der Richtungen gerade und *ausschließlich* darin begründet ist, daß die Linien sich *nicht* schneiden, bleibt unentschieden, so lange nicht der Satz bewiesen werden kann, daß zwei Linien von verschiedenen Richtungen (in der Ebene) sich schneiden *müssen*; und nicht nur dieser Beweis ist unmöglich, sondern der Satz selbst kann nicht einmal ausgesprochen werden, da wir ja über die Verschiedenheit der Richtungen erst durch das Auftreten des Schnittpunkts belehrt werden, so daß der Satz nichts sagen würde als: Linien die sich (endlos gedacht) schneiden, schneiden sich." (Gugler 1850, pp. 265–266).

[54] "In Kürze: Wir können *a priori*, ohne Rücksicht auf die Existenz eines Schnittpunkts, weder von Linien *gleicher* Richtung noch von Linien *verschiedener* Richtung sprechen, sondern wir haben zunächst blos Linien die sich schneiden und solche die sich nicht schneiden. Will man also durch die vorstehende Erklärung der Parallelen die Berufung auf die unendliche Verlängerung der geraden Linien vermeiden [. . .] so ist dies ein vergeblicher Versuch." (Gugler 1850, p. 266).

[55] "Die zweite (und wohl meist die entscheidende) Veranlassung zu jener Erklärung ist aber die schon oben angedeutete Rücksicht auf die Parallelentheorie. Hat man aus der Richtungsgleichheit zweier Geraden auf die Gleichheit der Winkel geschlossen, welche beim Schneiden durch eine dritte Gerade entstehen, so kann man allerdings hinterher *beweisen*, daß jene Geraden sich nicht schneiden. Allein wenn die unmittelbare Annahme gleicher Richtungen bei verschiedenen Linien schon an sich, ihrer Inhaltslosigkeit wegen, unerlaubt ist, so kann noch weniger erlaubt sein, Schlüsse von so weitgreifender Wichtigkeit darauf zu bauen." (Gugler 1850, p. 266).

Before I bring to an end this section on Frege let me emphasize that Frege's use of abstraction in *Grundlagen* introduces two novel elements with respect to the previous debates. The first, which I have already addressed at length, concerns the use of extensions as the value of the abstraction function. The second, which I have not yet introduced, is the use of higher-order abstraction principles. Frege is the first to work in this context with two sorts of entities, concepts and objects, and his discussion of the introduction of the number concept using one to one correspondence brings about a new situation with respect to the previous work. In particular, a higher-order abstraction sends concepts into objects and thus not all the possibilities that are represented in the taxonomy I gave at the end of chapter 1 can be deployed for interpreting this type of abstraction. According to that taxonomy, the three main options for assigning values to abstraction functions were 1) a choice of a representative from the equivalence class; 2) the equivalence class; 3) a new object not coinciding with the equivalence class or one of its representatives. With second-order abstraction we map concepts into objects. Since concepts are not objects possibility 1 is excluded as we do not take concepts as the value of the higher-order abstraction function. What about option 2? That will depend on the stand one takes as to whether Fregean extensions can be sets. My point of view is that there is enough evidence to think that Frege thought of his talk of extension to capture what mathematicians of his time called sets or manifolds. Consider the following passage from *Grundgesetzte*:

When logicians have long since spoken of the extension of a concept and mathematicians have spoken of sets, classes, and manifolds, then such a conversion forms the basis of this too [the transformation from the generality of an identity to an identity of value ranges]; for one may well take it that what mathematicians call a set, etc., is really nothing else but the extension of a concept, even if they are not always clearly aware of this. (Frege 1903, §147; tr. in Frege 2013, p. 148)

Should one resist the interpretation of Frege's extensions as sets, then one would have to claim that extensions are not equivalence classes of concepts and thus that Frege's extensions should be taken to be sui generis objects. This would affect both first-order abstractions, such as the one yielding directions, as well as second-order abstractions. In this case, we would be in the third class of our taxonomy. However, the third option sits well only with thinking of abstraction as yielding new objects that were not already at hand before (such as Dedekind's creation of the irrational numbers). It could perhaps be extended without too much trouble to Fregean first-order abstractions, such as directions, but it is definitely in tension with an outlook that postulates the full domain of entities as given from the outset. For instance, if the first-order quantifiers appearing in the right-hand side of Hume's Principle range over all the objects, then '$\#x:(Bx)$' for any concept B will be an object in the totality of objects *given at the outset* and thus in this sense it would not be a "new", sui generis, object that is added, so to speak, to a previously given domain of objects. I will come back on this issue at the end of Chapter 2.

We now move beyond Frege to the next stage of discussion of definition by abstraction.

2.2 The logical discussion on definitions by abstraction

2.2.1 Peano and his school

We have already seen that in 1887 Helmholtz referred to Grassmann in describing the process that leads from an equivalence relation to an identity of abstracta. His main examples came from physics. Of course, Frege in 1884 had also, but without using the terminology of abstraction, discussed the same type of concept formation. I have also shown in section 2.1 that Frege's §64 in the *Grundlagen* is influenced by Grassmann. Peano[56] (1888) contains the first description by Peano of definition 'by abstraction'. The terminology is not there yet (one has to wait to 1894 for the first full explicit use of 'definizione per astrazione' in a review by Vailati) but all the elements are in place. It is important to point out that the title of Peano's work is *Calcolo geometrico secondo l'Ausdehnungslehre di H. Grassmann*, which unequivocally shows Grassmann's influential role in shaping reflection on abstraction. In section 1, Peano defines equality between two entities of a certain system, written $a = b$, to mean a relation between elements of the system that satisfies symmetry and transitivity. Interestingly, he only states the properties but does not name them.[57] In section 80, Peano gives examples of relations that satisfy 1) neither symmetry nor transitivity; 2) symmetry but not transitivity; 3) transitivity but not symmetry; 4) both symmetry and transitivity. The latter are the important relations for definitions by abstraction. I recall that when a relation R satisfies that for *every* a there is a b such that aRb, reflexivity follows from symmetry and transitivity. His examples of relations that satisfy both symmetry and transitivity are:

1) the number a is equal to the number b

---

[56] On Peano and his school see Borga *et al.* (1985) and Roero (2010).

[57] The explicit use of notions such as reflexivity, symmetry and transitivity in the Peano school seems to originate with Vailati (1892) and De Amicis (1892). Vailati in 1892 claims originality for introducing the word 'reflexivity'. De Amicis also credits Vailati with the introduction of 'reflexivity' and both credit de Morgan with the introduction of 'transitivity'. De Amicis coined 'convertible' [*conversivo*] for what we call 'symmetric' but his terminology did not catch on. Symmetric, in this sense, was introduced by Schröder in 1890.

2) the number a is congruent to a number b with respect to a fixed module[58]
3) the straight line a is parallel to the straight line b
4) the straight line a coincides with the straight line b
5) figure a can be superposed over figure b

These are called equalities according to the definition given in section 1. The relation of identity is a case of equality but not every equality coincides with identity. Peano explains that one can define several equalities over a system of entities depending on the specific properties of the entities one decides to take into consideration.

Every equality between the entities of a system that is different from identity is equivalent to the identity between the entities that are obtained from those of the given system abstracting from all and only those properties that distinguish an entity from those equal to it. Thus, the equality 'the segment AB can be superposed onto the segment $A'B'$' is equivalent to the identity between entities that can be obtained from every segment by abstracting from all those properties that distinguish it from all those to which it can be superposed. The entity that results from this abstraction is called the *magnitude* [*grandezza*] of the segment; the former equality is thus equivalent to the identity of the magnitudes of the two segments. If we agree to indicate identity with the sign =, the equality just considered can be written as

$gr\ AB = gr\ A'B'$

Analogously, the equality 'the line AB is parallel to the line $A'B'$' can be written

direction AB = direction $A'B'$

and so on.[59]

---

[58] This is the only time that a number-theoretic example is given in this context by the Peano school. This occurrence does not contradict my claim that number-theoretic examples were not considered candidates for abstraction by Frege and the Peano school.

[59] "Ogni uguaglianza tra gli enti di un sistema, diversa dall'identità, equivale all'identità tra gli enti che si ottengono da quelli del sistema dato astraendo da tutte e sole quelle proprietà che distinguono un ente dai suoi eguali. Così l'eguaglianza 'il segmento AB è sovrapponibile al segmento $A'B'$' equivale all'identità tra gli enti che si ottengono da ogni segmento astraendo da tutte le proprietà che lo distinguono da quelli con cui è sovrapponibile. L'ente che risulta da questa astrazione viene chiamato *grandezza* del segmento; l'eguaglianza precedente equivale quindi all'identità delle grandezze dei due segmenti. Se conveniamo di indicare col segno = l'identità, l'eguaglianza ora considerata si potrà scrivere:

$grAB = grA'B'$

Analogamente, l'eguaglianza 'la retta AB è parallela ad $A'B'$' si può scrivere:

direzione AB = direzione $A'B'$

e così via." (Peano 1888, pp. 152–154).

Peano comes back to the topic of definitions by abstraction in 1894 in his "Notations de Logique Mathématique". After having discussed explicit definitions, in §38 he turns to a new sort of definition that is important in mathematics (there is no reference to Frege in Peano's discussion). He actually does not quite talk of 'definition by abstraction' but he claims that one introduces concepts by abstraction and by doing so one defines an equality.

There are concepts that are obtained by abstraction which constantly enrich the mathematical sciences but that cannot be defined in the stated form [namely, with an explicit definition, PM]. Let u be an object; by abstraction one obtains a new object ϕu; one cannot form an equality

$\phi u = $ known expression

since ϕu is an object whose nature is completely different from all those that have hitherto been considered. Hence one defines the equality by stating[60]

$h_{u,v}. \rightarrow: \phi u = \phi v. = .p_{u,v}$ Def.

where $h_{u,v}$ is the assumption on the objects u and v; $\phi u = \phi v$ is the equality that is being defined; it has the same meaning as $p_{u,v}$, which is a condition, or relation, between u and v, with a well known meaning. (Peano 1894, p. 45)[61]

The claim that the left-hand side and the right-hand side of the equivalence have the same meaning will be repeated by many members of the Peano school but it is of course something many people will dispute.[62] Then Peano states that the equality among the newly introduced objects, and a fortiori the equivalence relation, must satisfy the properties of reflexivity, symmetry, and transitivity (this time using this terminology; see also Burali-Forti 1894b). I would like to draw attention to an important shift between this presentation and the one in 1888.

---

[60] For typographical reasons, I have replaced throughout the Peano symbol for material conditional with →.

[61] "Il y a des idées qu'on obtient par abstraction, et dont s'enrichissent incessamment les sciences mathématiques, qu'on ne peut pas définir sous la forme énoncée. Soit u un objet; par abstraction on déduit un nouveau objet ϕu; on ne peut pas former une égalité

$\phi u = $ expression connue,

car ϕu est un objet de nature différente de tous ceux qu'on a jusqu'à présent considérés. Alors on définit l'égalité, et l'on pose

$h_{u,v}. \rightarrow: \phi u = \phi v. = .p_{u,v}$ Def.

où $h_{u,v}$ est l'hypothèse sur les objets u et v; $\phi u = \phi v$ est l'égalité qu'on définit; elle signifie la même chose que $p_{u,v}$, qui est une condition, ou relation, entre u et v, ayant une signification bien connue." (Peano 1894, p. 45).

[62] It is, for instance, denied by Crispin Wright and Bob Hale who however try to articulate in which sense the left-hand side and the right-hand side of the equivalence have the same content.

One of the examples given by Peano in 1888 was that of congruence in number theory. Here congruence is not mentioned among the examples. And this is certainly not on account of the fact that congruence in 1894 had ceased to satisfy the relevant properties for being an equivalence relation. Rather, the ontological spin given by Peano to definitions by abstraction, namely what is introduced is a *new* object ϕu, does not capture what happens in number theory where the object introduced is usually a representative of the equivalence class and thus not a new object.

Peano says that a relation that satisfies the three properties in question has 'the properties of equality'. The object denoted by ϕu is what one obtains considering all and only the properties that it has in common with all other objects v that are equivalent to u, so that one also has ϕv.

Peano then provides a list of examples that are by now familiar to us, beginning with the theory of ratios in Euclid. He also presents examples taken from arithmetic (broadly construed): the introduction of rationals from the integers by means of pairs using Stolz (1885), the introduction of irrationals as *lim sup* of sets of rationals. Moving to geometry, Peano mentions the introduction of length and direction. About the latter he writes:

The relation between two unbounded straight lines "a is parallel to b" has the properties of equality. It has been transformed into "direction of a = direction of b", or "point at infinity of a = point at infinity of b". One cannot define an equality of the form: "point at infinity of a" = "expression formed with the words of Euclid's Elements."[63]

The latter remark is doubly connected to the ontological role that Peano ascribes to definitions by abstraction. It is exactly because the entity is undefinable using the previous vocabulary that a definition by abstraction results in something ontologically fruitful. But once again, I repeat, this also shows why the normal use of abstraction in number theory is, from this point of view, spurious in that the entity introduced (the representative) is simply one of the old entities and thus definable if the old entity was definable or available as a primitive entity. By way of further geometrical examples, Peano mentions the introduction of vectors, quaternions as pairs of vectors, and concludes with Cantorian cardinalities, points beyond infinity in hyperbolic geometry, and Grassmann's geometrical forms. He also warns that the introduction of definition by abstraction is not always fruitful or desirable. He mentions as an unfruitful abstraction the case of 'shape' (arising

---

[63] "La relation entre deux droites illimitées "la a est parallèle à la b" a les propriétés de l'égalité. Elle a été transformée en "direction de a = direction de b", ou "point à l'infini de a = point à l'infini de b". On ne peut pas former une égalité de la forme: "point à l'infini de a" = "expression composée avec les mots des Éléments d'Euclide" " (Peano 1894, p. 47).

from similarity of geometric figures) and then states that projectivity also satisfies the conditions for introducing an equality.

In section 40, Peano comments on the fact that the equalities introduced by a definition by abstraction are true identities and explains why there are no failures of substitutivity (i.e. why one cannot argue from $2/3 = 4/6$ and $2/3$ is an irreducible fraction to $4/6$ is an irreducible fraction).

At this point, many people in the Peano school began writing about definitions by abstraction. I will also recall that Russell in 1900 discovered Peano's contributions and this led to his (Fregean) techniques of eliminating definitions by abstraction in terms of explicit definitions. But such techniques had already been anticipated by Burali-Forti, who will later repudiate them.

Let us consider Burali-Forti's *Logica Matematica* of 1894b.[64] Burali-Forti offers a taxonomy of definitions into four types and considers definitions by abstraction as the fourth type in his classification (pp. 140–145). The exposition is very similar to that given by Peano in 1894 although there are small variations in terminology. Burali-Forti explains that such definitions are used "when the entity x that one wants to define is obtained as an abstraction of a determined complex of known entities". The primary example discussed by Burali-Forti concerns the definitions of the abstract entities 'rational numbers'. Burali-Forti comments:

In some cases not even the previous type of definition [by postulates, PM] can be adopted. This happens when the entity x that one wants to define is obtained as an abstraction of a determined complex of known entities.

A rational number, for instance, can be defined as an abstract entity obtained from a pair of integers. If m, n are integers, with m/n we indicate an entity which depends on m and n and the mode of dependency, which can be stipulated at will (except for keeping in mind what result one wants to reach), defines, by abstraction, the entity m/n.

In general: if u is the known entity (for instance, the pair m, n of integers), then the thing that one wants to define (for instance, the rational number m/n), is a function ϕ of u. For ϕu one needs to define the relation indicated by the sign $=$, saying what is the meaning, for the things $\phi u, \phi v$, of the relation $\phi u = \phi v$. Such Def has the form

$$h_{u,v}. \rightarrow: \phi u = \phi v. = .p_{u,v}$$

where $h_{u,v}$ is the assumption relative to the things u, v; $p_{u,v}$ is the proposition, whose meaning is already known, containing u, v, and that we set equivalent to the relation $\phi u = \phi v$ that has to be defined.[65]

---

[64] Burali-Forti's *Logica Matematica*, in both the 1894 and 1919 editions, has been recently reprinted with an insightful introduction by Gabriele Lolli (see Burali-Forti 2013 and Lolli 2013).

[65] "In certi casi neanche la forma precedente di definizione può essere adottata. Ciò avviene quando l'ente x che si vuole definire, si ottiene come astrazione di un complesso determinato di enti noti. Un razionale, p. es., può esser definito come ente astratto ottenuto da una coppia di numeri interi. Essendo m, n due numeri interi, con m/n indichiamo un ente che dipende da m e da n, e il modo

Burali-Forti immediately goes on to mention that, starting from an appropriate relation, one can also define new abstractions on abstract entities. However, Burali-Forti does not seem to realize that this only defines an identity between terms of the form $\phi u = \phi v$ and not an arbitrary identity $\phi u = x$ (for an x which is not given in the form ϕv, for some v). Indeed, he goes on to assert that one can also provide an explicit definition of the class of entities :

The class H of entities defined as abstract functions ϕ of known entities $u, v, \ldots$ is defined, by means of a definition of *first species* [i.e. an explicit definition, PM], by setting

$$H = x\epsilon\,(u\epsilon M.x = \phi u.- =_u \Lambda)$$

where M is a known class, and this definition is read "H is the complex of entities x such that there exists at least a u in the class M for which it holds that x is *identical* to ϕu"[66]

Frege had given up definitions by abstraction because the equality $x = \phi y$, where x is not given in the form ϕz, for some z, is not specified by the definition by abstraction.[67] One should add that in the Fregean case the situation was more dire on account of the need to use such equalities in contexts in which they simply could not be eliminated.[68] Burali-Forti discusses in detail the introduction of rational and irrational numbers. He points out that the right-hand side in such definitions must be given by a relation which satisfies the properties of reflexivity, symmetry, and transitivity,—properties that are in turn inherited by the equality so that $\phi u = \phi u$, if $\phi u = \phi v$ then $\phi v = \phi u$, and if $\phi u = \phi v$ and $\phi v = \phi w$ then $\phi u = \phi w$. This is followed by some geometrical examples with equivalence of

di dipendenza, che noi possiamo stabilire ad arbitrio, (salvo il risultato al quale si vuol giungere), definisce, per astrazione, l'ente m/n. In generale: se u è la cosa nota (p.es., la coppia m, n di numeri interi), allora la cosa che si vuol definire (p. es. il razionale m/n), è una funzione ϕ di u. Per ϕu bisogna definire la relazione indicata dal segno $=$, dicendo quale è per le cose ϕu, ϕv, il significato della relazione $\phi u = \phi v$. Tale Def ha la forma

$$h_{u,v}. \to: \phi u = \phi v. = .p_{u,v}$$

ove $h_{u,v}$ è l'ipotesi relativa alle cose u, v; $p_{u,v}$ è la prop. contenente u, v avente già significato noto, e che poniamo equivalente alla relazione $\phi u = \phi v$ da definire." (Burali-Forti 1894b, p. 140).

[66] "La classe H di enti, definiti come funzioni astratte ϕ degli enti noti $u, v, \ldots$ risulta, con una definizione di *prima specie*, definita, ponendo

$$H = x\epsilon\,(u\epsilon M.x = \phi u.- =_u \Lambda)$$

ove M è una classe nota, e tale definizione si legge "H è il complesso degli enti x tali, che esiste almeno un u della classe M, per il quale x è *identico* a ϕu". " (Burali-Forti 1894b, p. 141).

[67] On account of one of the examples used by Frege in his discussion this has become known as the Caesar problem.

[68] The issue, which concerns Frege's definition of the successor through the ancestral, has been discussed at length in Dummett (1991), Hale and Wright (2001b), and Heck (2011), just to name three prominent examples from the extensive literature on this matter.

plane figures, parallelism of lines and planes (thereby introducing areas, directions or points at infinity, and orientation [*giacitura*]). I remark that the relation of congruence modulo a certain natural number is not given as an example (nor is any other example from number theory). Finally, an interesting example comes with the notion of 'meaning' [*significato*]. Burali Forti abstracts from logically equivalent propositions to obtain the 'meaning' or the 'value'[69]:

If A and B are propositions, the relation $A \to B.B \to A$ or, A is *equivalent* to B, is reflexive, symmetric, and transitive (p. 27). We obtain then from each proposition A the abstract entity *value of A* or meaning of A: and we say that "The meaning of A is *equal* to the meaning of B, just in case A is equivalent to B". An analogous observation holds when A and B are classes.[70]

When applied to classes, this principle is nothing else than Frege's Basic Law V (whether Burali-Forti had seen Frege's 1893 work at this stage, I do not know; Peano published a review of it in 1895). Finally, Burali-Forti explains that there is no failure of substitutivity originating from the definitions by abstraction under consideration. The examples are the same as those Peano used in 1894 concerning irreducible fractions.[71]

A perusal of Burali-Forti's work in set theory during this period shows that the terminology preferred by him is that of 'introduction of abstract entities'. Thus in 1896*a* in his work on finite classes, he presents the Cantorian introduction of cardinal numbers by means of a definition by abstraction and claims that the left-hand side and the right-hand side of the equivalence have the same meaning. The finite cardinalities are seen as 'abstract entities' which are the values of functions with domain the finite classes (see 1896*a*, note 1, p. 51).[72] In his 1896*b*, Cantorian cardinalities and ordinal numbers are also introduced by means of definitions by abstraction but without using this terminology and talking about the introduction of 'abstract entities' which are functions of given classes or, as Burali-Forti says

---

[69] Brandom (1986, p. 281 and p. 292) persuasively points out that all semantic notions in Frege (sense, reference, thought, truth value etc.) are introduced by means of definitions by abstraction.

[70] "Se A, B sono proposizioni, la relazione $A \to B.B \to A$ o, A è *equivalente* a B, è riflessiva, simmetrica e transitiva (p. 27). Otteniamo allora da ogni prop. A l'ente astratto *valore di A* o significato di A: e diciamo che "Il significato di A è *eguale* al significato di B, quando A è equivalente a B". Analoga osservazione vale quando A, B sono classi." (Burali-Forti 1894*b*, p. 147).

[71] "Così, p.es., 4/5 è *frazione irreduttibile*; ma $4/5 = 8/10$; dunque 8/10 è *frazione irreduttibile*. Il che è falso. Ed è falso perchè *frazione irreduttibile*, non è una funzione del razionale 4/5, ma invece una funzione della coppia $(4, 5)$, diversa dalla funzione R '$(4, 5)$ o 4/5. Se dunque definiamo il razionale come una funzione di una coppia di numeri, con i termini *frazione ireduttibile* non indichiamo più un razionale nel senso inteso prima." (Burali-Forti 1894*b*, p. 148).

[72] Incidentally, a detailed study of this article by Burali-Forti would repay detailed attention. He aims at proving the Peano axioms for arithmetic from the abstraction principle for finite Cantorian cardinalities using only the logical notions of class and correspondence.

in 1894a, p. 177, one can 'define a class of abstract entities' (ordinal numbers) for which 'the identity is defined' by the appropriate equivalence.

In this 1894a paper, Burali-Forti refers to Peano 1894 which shows Peano's leading role in these reflections (we have already seen that Peano had written about such matters already in 1888). As far as I have been able to establish, the first complete and explicit use of the expression 'definition by abstraction' appears in a review of Burali-Forti 1894b written by Vailati in 1894. In the review, Vailati says that "the Author [Burali-Forti] gives special attention to the particular case of the so-called definition by abstraction [*definizione per astrazione*]". From now on the terminology becomes standard. Peano will use it in several articles and contributions starting in 1899 (see Peano 1899a, p. 12; Peano 1899b, p. 135; 1900, p. 13, and then in many publications and reviews in 1901). Especially interesting is the definition given in the dictionary of mathematics where he shows awareness that a definition by abstraction does not define in isolation the terms flanking the equality on the left-hand side of the biconditional. He says:

Abstraction. In mathematical logic one calls "definition by abstraction" the definition of a function ϕx with the form: $\phi x = \phi y. = .$ (expression formed with previously given signs), that is one does not define the sign ϕx in isolation, but only the equality $\phi x = \phi y$.[73]

In order to complete the treatment of the pre-Russell phase, let us consider briefly Burali-Forti 1899a, 1901 and Peano 1901a, b.

Burali-Forti 1899a shows already from its title the preoccupation with a characterization of equality, which is carried out in terms of the by now familiar technique of introduction of new abstract objects through a function operating on elements that are related by an equivalence relation. Using the standard stock of examples (length, area, volume, shape, direction, orientation etc.) from geometry (once again, congruency modulo n is not used), Burali-Forti recaps the elements of the theory and the distinction between equality and identity. The article is of interest only because it contains the roots of an 'operator' approach to the introduction of new entities, such as the rational numbers, that Burali-Forti will in the following years try to present as an alternative way to introduce by an explicit definition the entities normally obtained through a definition by abstraction. Indeed, he claimed that his definitions of rational and irrational numbers are to be considered as explicit definitions if the notions of magnitude and correspondence

---

[73] "Astrazione. Dicesi in logica matematica "definizione per astrazione" la definizione di una funzione ϕx avente la forma: $\phi x = \phi y. = .$ (espressione composta coi segni precedenti), cioè non si definisce il segno isolato ϕx, ma solo l'uguaglianza $\phi x = \phi y$." (Peano 1901c, p. 7).

are granted.[74] Burali-Forti went on to claim that it is only when one cannot give such an explicit definition that another approach is necessary and lists the definitions of cardinal number, ordinal number, direction, orientation, length, areas, volumes, mass, and temperature as relevant cases. But rather than describing the usual definitions by abstraction, he proceeded to show how in effect one can obtain any definition by abstraction as a consequence of an explicit definition of the class of elements that constitutes the range of the abstracting function. The text is hard to parse because rather than defining the range through an equivalence relation on which the abstraction must be considered, he defines the range by means of the function itself.

Moving now to the essential cases of definition by abstraction, Burali-Forti explains that by saying that x is a length one claims that x is a simple element and spells out this condition by stating that for such an element the phrase 'y is an x' does not have any meaning. He articulates a linguistic distinction between *length* and *length of*. Suppose length(x) is the function given by the definition by abstraction. For a specific segment a, length(a) is a simple element called the length of a. Then Burali-Forti explains:

When we say, for instance, that x is a length, we mean to say that x is a *simple element* that is a function (for example) of a segment [this simply means that $x = $ length(a) for a segment a, PM], and this function [that is length(a), which is x] is common to all the segments that can be superimposed to each other. If a is a segment, the function of a being considered [namely length(a), that is x] is indicated by the phrase *length of a*. Thus, while the word length is a common noun, a class, the word *length of* is a **correspondence** between the *segments* and the class *length*. In the same way, *cardinal number* indicates a class and *cardinal number of* indicates a correspondence between the *classes* and the simple elements that are the elements of the class *cardinal number*.[75]

Thus, *length* stands for a class containing the different lengths of segments, and thus it is not a simple entity, whereas each length(a), for a segment a, is a simple element which is a member of *length*. In the case of cardinal numbers *Cardinal* is the class, thus a complex entity, of all simple elements that are values of functions

[74] The same type of 'operator' approach is defended, in contraposition to the introduction of rationals by abstraction, in Peano's 1901a. For some considerations on Burali-Forti's operator theory, which I cannot further discuss here, see Lolli 2013.

[75] "Lorsque nous disons, par exemple, que x est une longueur, nous entendons dire que x est un *élément simple* qui est fonction (par exemple) d'un segment, et cette fonction est commune à tous les segments superposables entre eux. Si a est un segment, la fonction considérée de a est indiquée par la phrase *longueur de a*. Donc, tandis que le mot longueur est un **nom commun**, une **classe**, le mot *longueur de* est une **correspondance** entre les *segments* et la classe *longueur*. De même, *nombre cardinal* indique une classe, *nombre cardinal des* indique une correspondance entre les *classes* et des éléments simples qui sont les éléments de la classe *nombre cardinal*." (Burali-Forti 1899a, pp. 257–258).

preserving equinumerosity. I will not get into the details of the kind of advantage Burali-Forti thought he had accomplished by making this move. He defined the notion of cardinal number as 'one of the correspondences f between classes and simple elements such that for any class u, the classes v for which $fv = fu$ are all, and the only, classes similar to u'. In his discussion he also realized that this definition was not unique on account of the fact that several functions could satisfy the relevant condition. The most important thing concerning this contribution is that Burali-Forti thinks of the individual cardinals (as opposed to the class *Cardinal*) as simple elements that cannot be further analyzed. This is in contrast to Russell's definition of each cardinal as a class of classes.

In 1900, Russell met Peano and some members of his school in Paris and, as is well known, this was a turning point in his intellectual career. At the meeting in Paris, Peano had talked about definitions and Burali-Forti, who could not be present, sent a contribution on the definitions of irrational numbers that was read by Couturat. Peano (1901a) has nothing to offer[76] on the matter of definition by abstraction but Burali-Forti's article (1901) goes back to the issue and is relevant for us. Burali-Forti claims to be able to classify all definitions into three sorts: nominal, by postulates, and by abstraction. From the classification he attempted to draw an important philosophical distinction between concepts and intuitions. His idea was that any x that is defined by a nominal (explicit) definition is a concept. Intuitions are those x's that can be given only through a definition by postulation or a definition by abstraction. It follows, according to Burali-Forti, that whether something is a concept is an absolute notion, whereas whether something is an intuition depends on the state of science. Indeed, notions that were introduced by postulation (as in axiomatic systems) or by abstraction might then be able to be explicitly defined at a later stage of research. The formal characterization of definition by abstraction given by Burali-Forti in this article is standard but he repeats that one can define nominally the range of an abstraction function. For instance *direction* is $\{x :$ there is a straight line a such that $x = direction$ of $a\}$. Notice that, as previously explained, there is a difference between *direction* and *direction of*. The former is a class made up of simple entities (the directions of a, for arbitrary segments a). On the basis of this opposition between concepts and intuitions, Burali-Forti classified the definitions of number given by Dedekind and Peano (by postulates) and by Cantor (by abstraction) all as intuitions. He then contrasted those definitions with his definition of natural number, which he claimed to be an

---

[76] Actually, in this paper (but this was added only in 1901 after the paper was delivered), Peano contrasts in a footnote (p. 286) the introduction of fractions by abstraction and the 'operator' approach he is championing.

explicit definition. He made the same claim for his theory of rational numbers and the integers. I will not enter the details of the alleged explicit definition given by Burali-Forti for it would force me to present his theory of magnitude. I will only mention that it is clear that we have here some sort of foundational program that aims at eliminating definitions by postulation and definitions by abstraction in favor of nominal (explicit) definitions. I think this goal was shared also by Peano and affects Russell's approach to this issue. Regardless whether one can grant Burali-Forti the successes he claimed, both Peano and Burali-Forti agreed that Cantor's theory of cardinal and ordinal numbers could not be reduced to explicit definitions at the then current state of science.

2.2.2 Russell and Couturat

The above was the state of the discussion on definition by abstraction when Russell discovered the Peano school and entered the fray with his article on relations, Russell (1901). Not surprisingly, we find Russell emphasizing the issue of when a definition by abstraction can be turned into a nominal definition. There is a draft of Russell (1901) published in vol. 3 of the *Collected Papers*[77] where Russell's first reference is to Burali-Forti (1901) (Russell 1993, p. 590). Even the first claim, in the original English version, about the advantage of introducing functions defining them through the notion of relation is strictly connected to the previous debate:

> The following notation appears to introduce at once a simplification and a generalization of many mathematical theories; and it enables us to render all definitions nominal. (Russell 1993, p. 590)[78]

Russell's contribution to abstraction has already been the subject of scholarly scrutiny and thus I will be brief.[79] Gregory Moore summarizes quite clearly the major results of the paper on relation vis-à-vis definition by abstraction:

> To obtain a definition of cardinal numbers, he [Russell] used what Peano called "definition by abstraction". That is, given an equivalence relation R, there is a function ϕx such that xRy if and only if $\phi x = \phi y$; thus, for example, the relation of one–one correspondence between two classes x and y gives rise to the function "cardinal number of x" (Peano 1894, 45). But Russell regarded it as necessary, in order to obtain such a function ϕ (or, as he preferred

---

[77] *Caveat lector*: the English translation published by Marsh and corrected by Russell in 1956 has important changes with respect to the original, in particular on the issues I am discussing.

[78] The 1956 translation reads like the French published text: "and it permits us to give nominal definitions whenever definitions are possible".(Russell 1993, p. 315) The reference to Burali-Forti, which was in the English draft (Russell 1993, p. 590), was however removed from the published version.

[79] See, for instance, the analysis of Russell on definitions by abstraction given in Vuillemin (1971), Rodriguez-Consuegra (1991), and Grattan-Guinness (2000).

to put it, a many–one relation S), to introduce a primitive proposition stating that any equivalence relation R can be written as the relative product of a many–one relation S and its converse (V.I, §I, *6·2). He applied this primitive proposition, which in the *Principles* he called the Principle of Abstraction (1903, 166), to the relation of similarity between classes, thereby obtaining such a relation S; then he defined the class Nc of all cardinal numbers as the codomain of S. But in a marginal comment by this definition, he recognized the problem that S is not uniquely determined: "This won't do: there may be many such relations as S. Nc must be indefinable" (V.1, §3, *1·4) While in his first draft (Appendix V.I) he freely referred to the cardinal number of a class without any mention of S, in the published version (Paper 8) he took a particular S as given and only defined individual cardinal numbers in terms of S. Nevertheless, sometime between February and July 1901, he added a sentence to the effect that, for any equivalence relation R, we can always take the equivalence class of a term u as "the individual indicated by the definition by abstraction; thus for example the cardinal number of a class u would be the class of classes similar to u" (8, 320). This was the famous Frege–Russell definition of cardinal number. Russell applied it not only to the relation of one–one correspondence, in order to obtain cardinal numbers, but to any equivalence relation whatever. (Russell 1993, p. xxvii)

In Russell's own words:

P6.2 is the converse of P6.1. It affirms that all relations which are transitive, symmetrical, and non-null can be analyzed as products of a many one relation and its converse, and the demonstration gives a way in which we are able to do this, without proving that there are not other ways of doing it. P6.2 is presupposed in the definitions by abstraction, and it shows that in general these definitions do not give a single individual but a class, since the class of relations S is not in general an element. For each relation S of this class, and for all terms x of R, there is an individual that the definition by abstraction indicates; but the other relations S of that class do not in general give the same individual. [. . .] Meanwhile, we can always take the class ρx, which appears in the definition of Prop 6.2, as the individual indicated by the definitions by abstraction; thus for example the cardinal number of a class u will be the class of classes similar to u. (Russell 1993, p. 320; [preliminary English draft of Russell 1901])

According to Russell then, we obtain in this way a nominal definition of the cardinal number of a class (in addition to the concept of cardinality itself). Indeed, in *Principles of Mathematics* (1903, p. 112) he will argue that the new theory of relations is the only one that allows to give up both definitions by postulation and by abstraction. Once again, he refers in a note to Burali-Forti (1901):

Moreover, of the three kinds of definitions admitted by Peano—the nominal definition, the definition by postulates, and the definition by abstraction [a note here refers to Burali-Forti 1901]—I recognize only the nominal: the others, it would seem, are only necessitated by Peano's refusal to regard relations as part of the fundamental apparatus of logic, and by his somewhat undue haste in regarding as an individual what is really a class. (Russell 1903, p. 112)

Like Burali-Forti, Russell worries about the non-uniqueness of the relation S (Burali-Forti's worry was the analogous one about the function corresponding to S). These latter issues will be central in the later discussion on Russell's transformation of definitions by abstraction into nominal definitions. The treatment of abstraction in *Principles* is somewhat confused due to the stratified nature of the composition of the text. In the middle of writing the book, Russell discovered his technique for eliminating abstraction but some passages in the text reflect an older state of things. This much he admitted in a letter, dated December 10, 1903, addressed to Couturat who had asked for clarifications on this matter (Russell 1993, pp. 726–727; Couturat's letter was dated December 7, 1903, see Schmid 2001, pp. 343–345).

In *Principles*, Russell proceeded, almost in Fregean manner, by first offering a definition of number by abstraction only to criticize it and replace it with a nominal definition.[80] Unlike Burali-Forti, Russell takes the values of the abstraction function to be classes and thus not simple entities. Peano in 1901b, p. 70, explicitly provides a definition by abstraction of Cantorian cardinalities, using the symbolism Num(a) for a class a, but explicitly rejects the idea of identifying Num(a) with the class of all classes that are in one–one correspondence with a, for he explains that Num(a) and the classes of classes that are in one–one correspondence with a are "objects which have different properties." It is unclear whether Peano's refusal to identify Num(a) with a class originates from a belief, shared with Burali-Forti, that Num(a) must be simple or whether purity of methods issues might be at stake.[81] I will recall here that when Dedekind discussed a proposal by Weber concerning a theory of number in which the numbers are defined as classes of classes (see Reck 2003, p. 385), he rejected the proposal exactly with motivations very similar to those of Peano's. Dedekind wrote:

If one wishes to pursue your approach I should advise not to take the class itself (the system of mutually similar systems) as the number (Anzahl, cardinal number), but rather something new (corresponding to this class), something the mind creates. (Cited in Reck 2003, p. 385)

Russell rejected Peano's position and claimed not to be able to see what these different properties between the two entities might be. He added:

---

[80] "I shall first set forth the definition of numbers by abstraction; I shall then point out formal defects in its definition, and replace it by a nominal definition." (Russell 1903, p. 112).

[81] Such objections are repeated in Burali-Forti (1909) and Catania (1911). According to Burali-Forti a "simple entity is one that is not a class."

Probably it appeared to him immediately evident that a number is not a class of classes. But something may be said to mitigate the appearance of paradox in this view. (Russell 1903, p. 115)

Russell adduced some considerations for making it less objectionable to identify numbers with classes and concluded that the strategy outlined for cardinal numbers could be used in all definitions by abstraction:

Wherever Mathematics derives a common property from a reflexive, symmetrical and transitive relation, all mathematical purposes of the supposed common property are completely served when it is replaced by the class of terms having the given relation to the given term; and this is precisely the case presented by cardinal numbers. For the future, therefore, I shall adhere to the above definition, since it is at once precise and adequate to all mathematical uses. (Russell 1903, p. 116)

Russell's main objection to definitions by abstraction consists in the fact that the function postulated in the abstraction is not unique:

Now this definition by abstraction, and generally the process employed in such definitions, suffers from an absolutely fatal formal defect: it does not show that only one object satisfies the definition. Thus instead of obtaining *one* common property of similar classes, which is *the* number of the classes in question, we obtain a class of such properties, with no means of deciding how many terms this class contains. (Russell 1903, p. 114)

In section 111 of *Principles* two possible solutions to the lack of uniqueness were discussed and eventually Russell settled for an identification of the number of a class *a* with the class of all classes that are in one-one correspondence [similar] to *a*. As we have seen, Russell proposed to apply this strategy to all definitions by abstraction.

Couturat (1905) also rejects definitions by abstraction in favor of explicit definitions obtained through Russell's principle of abstraction. After having explained what this principle amounts to, Couturat (1905, p. 50, note 1) uses a turn of phrase that stems from Russell's letter to Couturat from December 10, 1903: "Thus, the principle of abstraction does not lead to an abstraction but on the contrary it allows one to dispense with abstraction and to replace it"[82] (see also

---

[82] "Le principe d'abstraction n'a donc pas pour résultat d'effectuer l'abstraction, mais au contraire d'en dispenser et de la remplacer."(Couturat 1905, p. 50, note 1) Since the exchange between Russell and Couturat is not easily available I add the part of the correspondence that is relevant here. Couturat to Russell, December 7, 1903: "Seulement, je voudrais avoir un éclaircissement sur *le principe d'abstraction*. Vous dites (p. 166) que vous avez appliqué ce principe aux nombres cardinaux. Or dans la 2e partie je ne vois pas que vous ayez fait usage de ce principe, puisque vous y définissez le nombre cardinal comme une classe de classes (p. 115, 136). Et pourtant vous dites, dans votre *Préface* (p. IX), que le principe d'abstraction vous permet de définir les nombres comme classes. C'est ce que je ne comprends pas. Vous n'avez pas besoin de ce principe pour définir par ex. les classes *équivalentes*

Russell 1914). Needless to say, this method of elimination is also followed by Whitehead and Russell in *Principia Mathematica* even though the set up is much more complicated on account of the predicative theory of types motivated by the paradoxes.

Indeed, all would have been logically unobjectionable had it not been for the discovery of Russell's paradox. The discovery, which took place in summer 1901, i.e. in the middle of writing *Principles*, affected the reduction of definition by abstraction at least in those cases, such as cardinal (see section 111 of *Principles*) and ordinal (see section 231 of *Principles*) numbers, where the classes (of classes) turned out to be paradoxical. Russell already discussed this problem in *Principles*. On p. 305 he admitted that his method of turning definitions by abstraction into nominal definitions "is philosophically subject to the doubt resulting from the contradiction set forth in Part I, ch.x". In this connection, Russell referred to the appendix of *Principles* where he discussed at length Frege's system and the contradiction he had obtained from the axioms postulated by Frege.

We can now connect these developments to those citations of Weber (1906), Weyl (1910), and Hausdorff (1914) that we have already given in Chapter 1. In light

---

(similar); et ce principe peut vous server à déduire d'une classe de classes équivalentes l'idée du nombre cardinal qui est leur propriété commune. Il vous fournit donc les nombre cardinaux comme des entités singulières, et non comme des classes de classes. De même, quand vous l'appliquez à des *quantités égales*, il vous fournit la grandeur commune à toutes ces quantités, c. à d. *une et identique* en toutes. Il semble que vous ayez oublié d'appliquer votre principe au nombre, car vous ne le formulez explicitement que dans la 3e Partie (p. 166, 220) ce qui est un peu tard.

Autre question: Quelle est la valeur et la nature de ce principe? Ce n'est pas, apparemment, un principe premier, un axiome indémontrable (Pp.). Mais alors, comment le démontrez-vous? Cela me paraît difficile, car il est éminemment *synthétique*; en effet, il fait surgir d'une simple relation entre 2 entités une 3e entité nouvelle. Ecrivons en symbols:

$aSb. \rightarrow .aRc.bRc$

(*S* sym. et transitive; *R* rel. uniforme [many one]). On comprend la déduction inverse:

$aRc.bRc. \rightarrow .aSb(\text{car } S = RR)$

qui *élimine* c, mais la déduction directe, qui *introduit* c (et même le determine) paraît un peu forte, c. à d. paradoxale." (Schmid 2001, pp. 343–344)

Russell replied as follows on December 10, 1903:

"Au sujet du principe d'abstraction il se trouve sans doute une obscurité dans mon livre, qui resulterait de ce que je l'ai accepté autrefois comme axiome, tandis que je l'ai pu démontrer plus tard. La demonstration se trouve dans la Revue de Mathématique Vol. VII: je crois que le numéro est 6.28, mais je n'ai pas le volume ici. L'essentiel du principe, tel qu'il se démontre, est de substituer la classe même des objets dont il est question à la qualité hypothétique commune à tous ces objets. Au lieu de "principe d'abstraction" j'aurais mieux fait de l'appeler "principe remplaçant l'abstraction". Quand on a une relation *S* symmétrique et transitive, la classe des objets qui ont avec *a* la relation *S* remplace, dans le calcul, la propriété commune à tous ces objets, que suppose le sens commun. Je ne nie pas qu'il y ait souvent une telle propriété, mais il n'est pas nécessaire de l'introduire; elle serait en général indéfinissable, et la classe a toutes les qualités dont on a besoin." (Schmid 2001, p. 346).

of the paradoxes it was not at all clear whether definitions by abstraction could be dispensed with and Weber, Weyl, and Hausdorff were resigned to treat ordinal and cardinal numbers as objects introduced by abstraction about whose nature nothing more precise could be said.

The story of the various proposals addressing this paradox and the consequences for abstraction principles would have to take on board most of the debate on the foundations of mathematics between *Principles* and the formalization of most systems of logic and set theory well into the thirties. This is obviously not feasible here and the last section will instead look at the reactions within the Peano school to the Russellian solution and to the problem of accounting for definitions by abstraction.

2.2.3 Padoa on definitions by abstraction and further developments

One of the best papers written on definitions by abstraction is Padoa (1908). Padoa starts with explicit definitions (which is his preferred terminology for what Peano and Burali-Forti called nominal definitions). In an explicit definition one sets a (definitional) equality between expressions meant to indicate that the new expression is introduced as short-hand for a longer known expression. In this way the definiens acquires a meaning and the equality can be interpreted as 'means'. He then considered definitions by abstraction given in the following form. For K a known class and R an equivalence relation, the form of any definition by abstraction is:

If "a and c are (individuals belonging to) K" then "$Fa = Fb$" means aRb

where F is the function one wants to define. For instance:

if "a and b are polygons", then "area of a = area of b" means a is equivalent to b.

By keeping K as the class of polygons and letting R denote similarity, respectively equivalence, one defines the functions 'form of' and 'area of'. He added:

Since the *defined* notation is "$Fa = Fb$", equation 1 does not *explicitly* indicate either the meaning of F or that of "Fa". In other words, it only *authorizes* the use henceforth of the *sole* expression "$Fa = Fb$", because (using the reverse procedure) it teaches to substitute only to this expression an expression of known meaning ("aRb").[83]

---

[83] "Poichè la notazione *definita* è "$Fa = Fb$" la 1 non indica *esplicitamente* il significato nè di F nè di "Fa". In altri termini, essa *autorizza* ad usar poi la funzione F nella *sola* scrittura "$Fa = Fb$", perchè (usando il procedimento inverso) solo a questa essa insegna a sostituire una scrittura di significato noto ("aRb"). " (Padoa 1908, p. 94).

To begin with, Padoa entertains the possibility of considering that the definition by abstraction results in an explicit definition of the complex expression "$Fa = Fb$". He then asks whether this definition is arbitrary and points out that the question might appear paradoxical since all nominal definitions are arbitrary. Padoa argues as follows. In an explicit definition, say α, we have as a part of α the definiendum that contains new vocabulary with respect to the language available previously to introducing α. It would seem reasonable therefore that, using the old language and theory available before the introduction of the explicit definition α, α cannot be deduced from the theory. Padoa then rhetorically asks: would it not be reasonable also to say that α cannot be contradicted in the theory available before its introduction? But, he claims, this is not so, when the defined notion expresses a condition whose principal symbols are already known.

In order to explain his point, Padoa focuses on a relation R defined on objects of a class K. R is assumed reflexive, symmetric and transitive. He calls such a relation 'egualiforme' (I will translate this as 'equiform').[84] Let me remark that this is the first time that equivalence relations receive a special name to characterize them.

Now Padoa proves Theorem I, namely from the definition by abstraction

(1) if a and b are in K, "$Fa = Fb$" means "aRb", one infers
(2) R is 'equiform' in K

The argument, which relies on the assumption that the meaning of '=' on the left-hand side of (1) is known, shows that (1) cannot be assumed arbitrarily unless (2) has already been established. In other words, the 'equiformity' of R is a necessary condition for introducing a definition such as (1). Thus, if from the theory previous to the introduction of (1), the negation of (2) were to result, (1) would lead to contradiction; by contrast, if the previous theory does not allow either to prove or to refute (2), then (1) plays both the role of a definition and of a postulate.

Having thus established that definitions such as (1) are not arbitrary, taken as an explicit definition of the complex "$Fa = Fb$" when R is assumed "equiform", leads Padoa to ask whether (1) might not be interpreted as an implicit definition of F. And here the main objection to using F only in contexts such as "$Fa = Fb$" does not originate, says Padoa, from purely formal considerations, but from the fact that (1) does not individuate the meaning [il significato] of F. Padoa gives many examples of this indeterminacy of meaning. Let's consider one of them. If a and

---

[84] Even in the Italian contributions, this terminology will not assert itself. Burali-Forti (1912) speaks of a 'normal' relation and Cipolla (1914) of 'uniform' relation. Cipolla (1914) follows the account of definition by abstraction given by Russell and Padoa.

b are polygons, aRb means "a is equivalent to b" (a and b can be decomposed in parts that are respectively superposable) and "area" is the intended F given in (1). He then remarks:

We do not contest that (1) can be *legitimately* assumed as an *explicit Df* of "area of a = area of b"; but we deny that (1) *individuates* the *meaning* of "area of a".[85]

If the recipient had ignored the meaning of "Fa" as "area of", the definition by abstraction won't succeed in conveying it to him/her. In fact, one could attribute to F the meaning "twice the area of", "three times the area of", "half the area of", etc. (where 'area' is still taken in the original informal meaning). In short "area of" as it appears in a definition of the form (1) can be interpreted with infinitely many different meanings. One interesting objection to this line of thought that Padoa considers consists in claiming that the meaning of "area of" given by (1) has a meaning that consists of "all the meaning compatible with (1)" And here Padoa's objection does not seem so strong, for he says:

"But then before (1) can teach us the meaning of "area of" one needs to have found *all* its meanings compatible with (1); and how can the reader verify to have considered *all* of them?"[86]

How would, Padoa continues, the reader have included among the meanings of "area of a" that given by "volume of the prism having a base of area a and height a segment of constant length", which also satisfies (1)?

The point is repeated using examples with directions, orientations, etc. A useful footnote makes clear that in the case of directions one must use the relation "a is parallel to b or a coincides with b" which is an 'equiform relation' whereas 'a is parallel to b' is not. This is connected to some remarks I made in Chapter 1, namely that for much of the nineteenth century 'x is parallel to y' was not taken to be reflexive (which also implies failure of transitivity, although this was probably not always clearly seen).

So far the treatment has been critical. Now Padoa moves to a *pars costruens* and defines explicitly the "abstraction of a with respect to R" and proves a relevant theorem about it. Here is the passage:

Let us stipulate first of all the following *explicit Df*: **Definition**. If K is a *class* and R is an *equiform relation* on K and if a is an arbitrary element of K, then

---

[85] "Non contestiamo che la (1) possa *legittimamente assumersi* quale *Df esplicita* di "area di a = area di b"; ma *neghiamo* che la (1) *individui* il *significato* di "area di a" " (Padoa 1908, p. 97).

[86] "Ma allora, prima che la 1) ci apprenda il significato di "area di" bisogna aver trovato *tutti* i suoi significati compatibili con la 1); e come si accerta il lettore di averli considerati *tutti*?" (Padoa 1908, p. 98).

"**abstraction** of *a* with respect to *R*"

means "the set of all and only those *K* that bear the relation *R* to *a*."

For instance: if *a* is a *polygon*, then "*abstraction* of *a* with respect to the relation *is equivalent to a*" is "the set of all and only those polygons *x* such that *x* is equivalent to *a*".

Then we show the following

Theorem II. If *K* is a *class* on which *R* defines an *equiform relation* and if *a* and *b* are in *K*, then (indicating, for the sake of brevity, with F*a* and F*b* the *abstractions* of *a* and *b* with respect to R): "F*a* =F*b*" if and only if "*aRb*".[87]

Of course, this is the Russellian solution and one wonders how Padoa, after the publication of Russell's *Principles* in 1903 and of Couturat (1905), could present this result without attributing it to Russell. Perhaps the answer is at the very beginning of the article where Padoa says that his reflections go back to 1901 and that what he had in mind to present at a conference in Livorno (at the Second Congress of the Italian Philosophical Society) at the time (1901) was later developed by Russell and Couturat. But since, he continued, their writings had not yet enjoyed a large readership and because they did not treat the matter with the simplicity and generality required, he decided to write this article for the occasion of the meeting in 1906 from which the 1908 publication stems.

If theorem I showed that the relation *R* had to be 'equiform' as a necessary condition for the definition to be successful, theorem II provided a sufficient condition for it. Thus, the definition of "abstraction of" and theorem II constitute a "theory of abstraction". If one from the outset chooses, among the possible meanings for F, that entity given by "abstraction of", it turns out that a definition by abstraction of the form (1) is a theorem of the theory of abstraction. Coming back to the examples with areas of polygons we immediately reach an explicit definition of "area of *a*", for a polygon *a*, by setting "area of *a*" as "the abstraction of *a* with respect to the relation 'is equivalent to' ". Then the original definition by abstraction can be regained as a theorem of the theory of abstraction. Incidentally, also in Padoa there is no mention of Frege's work in this connection.

---

[87] "Stabiliamo anzitutto la seguente *Df esplicita*:

Definizione. Se *K* è una *classe* in cui *R* è una *relazione egualiforme* e se *a* è un individuo arbitrario di *K*, allora

"**astrazione** di *a* rispetto ad R"

significa "l'insieme di tutti e soli quei *K* che stanno nella relazione *R* con *a*." Ad es.: se *a* è un *poligono*, allora "*astrazione* di *a* rispetto alla relazione è *equivalente a*" è "l'insieme di tutti e soli i poligoni *x* tali che *x* è equivalente ad *a*".

Poi dimostriamo il seguente

Teorema II. Se *K* è una *classe* in cui *R* è una relazione *egualiforme* e se *a* e *b* sono *K*, allora (per brevità, indicando ordinatamente con F*a* ed F*b* le *astrazioni* di *a* e di *b* rispetto ad *R*): "F*a* = F*b*" quando e sol quando "*aRb*". " (Padoa 1908, p. 100).

Padoa also remarked that one can simply get rid of the word "abstraction" and simply define "area of a" as "the set of polygons that are equivalent to a". We are now moving towards the standard set-theoretic territory. But Padoa does not seem to have yet digested the lesson of Russell's paradox, for he also applies the reasoning to defining the number of a finite class a as the class of all classes that are in one–one correspondence with a.

Padoa concluded with the following methodological lesson:

One could object, I will not deny it, that common sense and experience will guide each time the writer to use definitions by abstraction exclusively when dealing with *equiform relations* on the considered class, thus avoiding any danger of contradiction. But it might have been useful to point out that if theorem I clarifies the *necessity* of such a *condition*, theorem II specifies its *sufficiency*.

And although the aid, real or alleged, of intuition would lead each time the writer to consider the concepts *ambiguously* defined *by abstraction* (in the usual way) as perfectly individuated, it will not have been useless to underline that the ambiguity was intrinsic to the *type* of D*f* [1] and to indicate a general procedure that—freeing us from the treacherous aid of *intuition*—will transfer the *explicit individuation* of *mathematical abstractions* from the psychological and epistemological realms to the *logical* realm.[88]

This last statement is a wonderful summary of the removal of the psychological roots of the abstraction process in favor of a merely logical description of it. It is exactly this move that bothered Angelelli (see 1984, 2004, 2013) but then again it seems to me that most of his worries reduce to an issue of semantics. Is it appropriate to refer to this definitional method with the word abstraction given that the tie with the psychological process has been severed? Perhaps not but what difference does it make? Let's call the logico-mathematical process abstraction*.

On the contributions that follow this paper by Padoa, I will have to be very brief. An important text in this connection is *Elementi di Calcolo Vettoriale, con numerose applicazioni alla Geometria, alla Meccanica e alla Fisica-Matematica*, by Burali-Forti e Marcolongo. The text was translated into French in 1910. It is important for two reasons. The first is that Burali-Forti attempts to bypass the use

---

[88] "Si potrà obbiettare, ed io non lo contesto, che il buon senso e l'esperienza guidano volta a volta il trattatista a giovarsi delle consuete D*f per astrazione* sol quando si tratti di *relazioni egualiformi* nella *classe* considerata, evitando così ogni pericolo di contraddizioni.

Ma può essere stato utile notare che, se il teorema I chiarisce la *necessità* di tale *condizione*, il teorema II ne precisa la *sufficienza*.

Ed ancorchè il soccorso, reale o presunto, dell'intuizione inducesse volta a volta il trattatista a ritener perfettamente *individuate* le idee *ambiguamente* definite *per astrazione* (nel modo consueto)—non sarà stato inutile rilevare che l'ambiguità—era insita a quel *tipo* di D*f* [1] e additare un procedimento generale che—affrancando dal malfido ausilio dell'*intuizione*—trasporti, dai campi psicologico e gnoseologico al campo *logico*, la *individuazione esplicita* delle *astrazioni matematiche*." (Padoa 1908, p. 103).

of Russell's classes of classes (against which he argues in favor of considering the abstract entities as simple) and does so by stating a "logical postulate" according to which the function F and the class that constitutes the range of this function are unique. This proposal was shown to be incoherent in Maccaferri (1913) and even though Burali-Forti (1912) already attempted to rectify his claim the only thing to point out about this latter article is Burali-Forti's rejection of definitions by abstraction in favor of a new operator theory that Burali-Forti will develop in the second edition of *Mathematical Logic* in 1919.

Incidentally, Maccaferri (1913) uses new examples to show the indeterminacy of meaning that is constitutive of definitions by abstraction but nothing he presents marks a decisive improvement on Padoa (1908), which Maccaferri discovered, as he declares in an afterword appended to the article, only after writing his article and with which he is pleased to agree. In his treatment Maccaferri ends up favoring the Russellian solution for turning definitions by abstraction into explicit definitions (or nominal definitions). He claims that nothing is simpler than choosing the Russell class as the value of the abstraction function and asks:

"Which function, of all the elements u related by the relation α, is *simplest* than the one that yields the very class of all those elements given that there is no criterion for choosing among them a unique element rather than any other one? And moreover every element b of the Russell class depends *only* on the elements that make it up, namely on the u's that are related among them by the relation α."[89]

The other reason why Burali-Forti and Marcolongo is of interest is for their lengthy appendix on definitions by abstraction that contains also the interesting definition by abstraction of Grassmann's forms. (See appendix to this chapter where the text is translated from the French edition).

Finally, I will mention Catania (1911) as an interesting discussion of the advantages and disadvantages of the Russellian principle of abstraction in comparison to that of Peano using only simple entities. Catania sides against Russell's approach and expresses optimism about the operator approach that Burali-Forti is developing (Catania himself will publish further work in this direction in the following years).

In Bindoni (1912) one finds an argument, resting on Burali-Forti's mistaken postulate, that the entities defined by abstraction and the class defined using

---

[89] "Quale funzione, di tutti gli u legati dalla relazione α, *più semplice* che non sia quella che fa ottenere la classe stessa di quegli elementi, poichè non c'è un criterio per scegliere tra essi l'uno elemento piuttosto che l'altro? E d'altra parte ciascun elemento b della classe di Russell dipende *soltanto* dagli elementi che lo compongono, cioè dagli u legati fra loro dalla relazione α." (Maccaferri 1913, p. 170).

nominal definitions (Russell-style) are identical. As for Peano 1911, it surprisingly contains no comments on definitions by abstraction. Peano 1915 however devotes a whole article to definitions by abstraction where he takes a rather pragmatic attitude as to which types of definitions are better. There are some interesting further examples of definition by abstraction relating to works by Fano on special relations between real numbers (such as 'belongs to the same algebraic field' or 'having a form $R \pm \sqrt{R'}$') and different orders of infinities (Mago 1913). Nothing Peano says in this article adds much to the previous discussion. However, he did claim credit for the expression 'définition par abstraction' by referring to his Peano (1894). But we have seen that this is only partially correct.

2.3 Conclusion

The foundational discussion on definitions by abstraction goes on to include Burali-Forti (1919), Weyl (1927), Carnap (1929), Natucci (1929), Dubislav (1931, 3rd edition), and Scholz and Schweitzer (1935) (see also Cassina 1961). The discussion in the 1920s and 1930s takes place against the background of the monumental *Principia Mathematica* and the use of definitions in context which is prominent in it. This naturally brings about a reconfiguration of the debate on definitions by abstraction but it is my sense that these works are mostly derivative from the previous discussion involving the Italian scholars centered around the Peano school, Russell, and Couturat.[90] In particular, the important ontological and semantical issues related to definitions by abstraction had already been characterized within the early discussion in the Italian school. Those discussions included:

a) ontological issues (the nature of the entities obtained by abstraction, simple or complex?, and, if complex, can they be identified with classes?);

b) semantical issues (indeterminacy of meaning and reference for the concepts and terms, respectively, obtained by abstraction; sameness of meaning between the left-hand side and the right-hand side of the equivalence);

c) logical issues (nature of the definition; necessary and sufficient properties required for its success; elimination of such definitions etc.).

Some of these points were of course foreshadowed by Frege and some were to reappear with a vengeance in the contemporary debate on neo-logicism.

It might be appropriate in the conclusion to this chapter to contrast the use of definitions by abstraction, as we have encountered them so far, and their use

---

[90] Vuillemin (1971) contains many interesting developments including discussions of Whitehead, Carnap, Russell, etc.

in contemporary neo-logicism. This might also serve as a bridge to the next two chapters.

There are of course some obvious differences between the use of abstraction principles in neo-logicism and the uses of abstraction we have encountered before Frege. Most obviously, no one before Frege entertained the idea of using definitions by abstraction as abstraction principles in the service of a philosophical program such as logicism. Frege himself entertained the possibility in the realm of arithmetic but then discarded it on account of the Caesar problem, namely the problem of how a definition by abstraction can provide a way to settle mixed identities when one of the terms flanking the identity does not have the form 'the number of *C*'. This problem appeared insurmountable to Frege and thus he was led to introduce extensions. In the later *Grundgesetze*, extensions are seen as special cases of courses of values and some of the problems raised by Frege in *Grundlagen* return with full force (see Brandom 1986 and Heck 2011) although in this case it was the contradiction which Basic Law V gave rise to that led Frege to despair. Contemporary neo-logicists do not share Frege's negative assessment, expressed in *Grundlagen*, of the prospects for introducing concepts (such as the number concept) by means of abstraction principles and have deployed a number of strategies for giving a solution to the Caesar problem. Presenting or evaluating these strategies won't be necessary for our goals.

Notwithstanding the above, following up on the discussion of Frege at the end of section 2.1, I would like to discuss the relationship between nineteenth-century uses of abstraction and contemporary uses in neo-logicism at a more basic level, namely that of the interpretation of abstracta yielded by abstraction principles. At the end of the first part of this chapter, I mentioned the novelty that higher-order abstraction represents with respect to the previous tradition. One can already appreciate the particular spin given by the neo-logicists to abstraction principles by considering first-order abstractions.

In Wright (1983), the relevant discussion starts with the definition of a sortal concept. On p. 2 we read:

Let us say that a concept is sortal if to instantiate it is to exemplify a certain general kind of object—not necessarily a natural kind—which the world contains.

Paradigm examples of what the above definition aims at capturing are concepts such as 'person', 'chair', 'duck' and 'helm'. Paradigm examples of what the definition intends to exclude are 'heavy', 'brown', 'bald' and 'striped'.

According to Dummett and Wright, defining sortal concepts requires both a *criterion of application* and a *criterion of identity*. The criterion of application is what allows one to discriminate between objects that fall under a certain concept from those that do not. The criterion of identity allows us to determine for any two objects that fall under the relevant concept whether they are identical or not. The former is a prerequisite for grasping any concept but the latter is more specific to sortal concepts.

The key to understanding the neo-logicist appeal to sortal concepts is to focus on the aim of the neo-logicist, namely to guarantee through abstraction principles the intelligibility of abstract sortal concepts, where the latter are defined as sortal concepts whose instances are abstract objects.

It is for this reason that the first option we have encountered in our historical discussion, namely choosing representatives of equivalence classes, is of no real interest or use to the neo-logicist. Consider, for instance, Wright's discussion of direction in Wright (1983). What Wright is concerned about is to show that an abstract sortal concept of direction can be gained, despite empiricist or nominalist doubts, through the abstraction

$$\text{Dir}(a) = \text{Dir}(b) \text{ iff } a//b$$

Wright starts with a and b as spatio-temporally located lines. That lines are not taken as abstract geometrical objects is part of the dialectic of the situation, for the goal is to show how abstraction principles can give us access to an ontological reality of abstract objects, contrary to empiricist reservations. For this reason, the neo-logicist has no interest in those uses of abstraction in which the function defined by the abstraction selects representatives from equivalence classes. Starting from spatio-temporally located objects such abstraction would only yield spatio-temporally located objects and thus would turn out to be unfruitful for the goals of the neo-logicists. However, notice that this possibility opens the way for an empiricist/nominalist to accept a non-austere interpretation (see Chapter 1, note 29) of the left-hand side of the abstraction principle while refusing to accept abstract objects. In other words, the non-austere empiricist/nominalist can take the functional terms flanking the equality of the left-hand side to behave syntactically and semantically as terms, and thus to denote, but insist that they denote objects from the original domain (which for the empiricist/nominalist will consist only of spatio-temporally given objects). Thus, in the example treated by Wright, the empiricist would not be forced to ascend to a reality of abstract objects but could simply, for a spatio-temporally given line a, reinterpret $\text{Dir}(a)$ as some line b from the original domain. By a simple logical result, this option is always available for first-order abstractions (one needs of course to make use of the axiom of choice). This position is what Heck in some of his articles calls *semantic reductionism* (see Heck 2011, chapters 8 and 9).

More promising for the neo-logicist aims are the forms of abstraction that give either equivalence classes (or extensions/courses of values in Frege's framework) or sui generis objects not identifiable with previously available spatio-temporally located objects. Both solutions can be interpreted as possible routes for yielding abstract objects as required by the neo-logicist. However, neither interpretation will persuade the committed empiricist/nominalist (whether of the austere or of the semantical reductionist variety). The first technique is hostage, if it has to be used for purposes the neo-logicist has in mind, to the precise characterization of the nature of an equivalence class, or an extension (or a course of values) in Frege's sense. Consider the case of extensions. If extensions are defined by astraction, for instance by assuming something along the lines of Basic Law V or a modification thereof, they will be objects of the domain given at the outset. But if that domain only contained spatio-temporally located objects, which is all the empiricist/nominalist will commit to, the extension of concepts would also turn out to be a spatio-temporally located object. This argument could also be applied to the case of directions but in the case of directions the way out for the neo-logicist is that of considering directions as sui generis objects not identifiable with previously available spatio-temporally located objects, that is, the spatio-temporal lines from which Wright takes his start. Indeed, it was Wright's strategy to argue, against the empiricist/nominalist, that one could successfully reach the realm of abstracta denoted by terms for directions introduced through the abstraction principle for directions. But the line of thought we used for extensions, which can be extended to Hume's Principle and other second-order abstractions, leads to a more serious worry for the introduction of the concept of number as an abstract sortal concept. An argument analogous to the one used in going from spatio-temporal lines to abstract directions cannot be applied in this context. For suppose the domain of objects is given at the outset and, in analogy with the case for directions, this domain contains only spatio-temporal entities (let's also assume there are infinitely many). When we define $\#(C) = \#(D)$ iff $C \sim D$ all the first order quantifiers are ranging over spatio-temporal entities and thus, since $\#$ sends concepts into objects, $\#(C)$ will be a spatio-temporal object.

Perhaps, one aspect of the Caesar problem is exactly this. How does one persuade the committed empiricist/nominalist, without begging the question at the outset by postulating the existence of abstract entities in the domain of quantification, that the denotata of terms such as $\#(C)$ are abstract objects as opposed to spatio-temporal objects (such as Ceasar)? Whatever the prospects for the neo-logicist, we do notice a disanalogy between the case of directions and that of number. In the case of second-order abstractions, such as Hume's principle, one must already assume that the abstract objects were already part of the original domain of quantification, for a strategy similar to the one presented by Wright in the case of directions would get us nowhere.

2.4 Appendix

Translated from Burali-Forti, C., and Marcolongo, R., *Éléments de calcul vectoriel avec de nombreuses application à la géométrie, à la mécanique et à la physique mathématique*, Hermann, Paris, 1910, pp. 213–216.
Historical and Critical Notes.
Note I
On definitions by abstraction
It is well known that, for instance, prime number and spherical surface can be given a definition (nominal or absolute) of the following form:

"prime number" = "integer with only two divisors"
"spherical surface" = "locus of points that are equidistant from a given point".

These are definitions in which the right-hand side has a *well known* and *precise* meaning.
The same method of definition cannot be applied to *vector, geometrical forms, direction, length, weight, mass*, etc. For such entities it is better to employ a type of definition that can be called *definition by abstraction*. We deem it useful to state explicitly in what it consists and how one must apply it, for one often applies it in an imprecise fashion.
Equality. One can give, following Leibniz, an *absolute* and *universal* meaning to the sign $=$.

(1)   "if x and y are arbitrary entities, one has $x = y$ if and only if *every property of x is also a property of y*".

And since "to be a property of x" can always be logically expressed by the statement "x is an element of a certain class u", the previous statement takes the form

(1')   "$x = y$ if and only if *every class u that contains x also contains y*".

Let us immediately make the following remark, which we will use later on: once we have defined in this fashion, following Leibniz, the relation designated by the symbol $=$, it is no longer allowed to define once more, for some newly introduced elements, the relation $x = y$.
From (1) and (1') one infers

(2) $\begin{cases} x = x \\ x = z \text{ and } y = z \text{ entail } x = y \text{ (Euclid)} \end{cases}$

and it is to these two properties that are reduced the *three* usual properties, reflexive $(x = x)$, symmetric $(x = y$ entails $y = x)$ and transitive $(x = y$ and $y = z$ entail $x = z)$. It is customary to say that the properties given in (2) are the *characteristic properties of equality*. This is not exact. In fact, the relations "it is equivalent to", "is similar to", for instance, satisfy the conditions in (2). However, they do not satisfy condition (1) because the relation "x is equivalent to y and y is a triangle" does not necessarily entail "x is a triangle". And similarly "x is similar to y and the volume of y is a cubic meter" does not necessarily entail "the volume of x is a cubic meter". In other words, condition (1) entails the conditions in (2) but the converse does not hold. While (1) defines a unique relation, *identity* or *absolute equality*, the properties in (2) define a class of relations, and the number of these relations is infinite, for we cannot put a bound to their construction. It follows that for the infinite relations defined by using (2), *it is not allowed to make use of the sign $=$ if one wants to preserve this sign to designate, as it is customary, the precise Leibnizian sense of absolute equality or identity.*

Definition by abstraction. This type of definition is based on the following principle of general logic.

Let us assume that for whichever elements x, y, z of a class u, the relation α between the [elements of] u, satisfies the following properties:

(3) $\begin{cases} x\alpha x \\ x\alpha z \text{ and } y\alpha z \text{ entail } x\alpha y; \end{cases}$

Then there exists a unique class v and a unique function f which satisfies the following properties.

1. *For any x in u, fx is an element of v;*
2. *For any h in v, there exists at least one element x of u such that $fx = h$;*
3. *For any x and y in u, one has $fx = fy$ if and only if x is in the relation α with y, that is $x\alpha y$.*

It follows that v and f are determinate functions of u and α; the proposition $x\alpha y$ between the pairs x and y, infinite in number, of elements in u, is expressed by means of the unique identity $fx = fy$; for the infinite number of pairs of elements of u related by α, one can substitute a unique element of the class v.[91]

Let us apply for instance definition by abstraction in order to obtain the geometrical entities called *directions* (or *points at infinity*). The class u will be the class of *straight lines*; α will express the relation "*is parallel to*". Since the conditions in (3) are verified[92], the function f exists: we will call it *direction*. In what follows the class v of *directions* also exists.

In a similar way one obtains, starting from the relation "is parallel to" for planes, the class of *directions of planes* (or *lines at infinity*); from the relation "it can be superposed over" for segments, one obtains the *lengths*; from "is equivalent to" the *areas* and *volumes*; from "is similar to" the *shape* of a geometrical figure. And starting from suitably chosen relations one obtains the *weights, masses, temperatures, quantities of heat,..., the sense of a succession of 2, 3, 4 points* etc.

---

[91] This last remark shows the practical importance of definition by abstraction which yields for instance the rational numbers, the real numbers, the complex numbers, as simple entities whereas for Mr. Russell they are *classes, classes of classes, ordered pairs of classes of classes*, which are very complicated to handle.

[92] The line a is parallel to the line b in the following two cases: 1) if a coincides with b; 2) if a and b are on the same plane and have no point in common. This is the definition that we adopt here and not Euclid's definition.

The *proper* definition (from a logical point of view) of *vectors* that we should have been given [this refers to the textbook from which this appendix is taken, PM] is the following.

Let us consider the class u containing all the pairs of points and the relation α defined as follows:

$(A, B)\alpha(C, D)$

if and only if

(4)   "the middle point between A and D" = "middle point between B and C".

The relation (α) verifies the conditions (3). Consequently, the function f and the class v exist. Let us write, by definition,

$f(A, B) = B - A$

and let us call "vector from A to B" the entity $f(A, B)$, or $B - A$; In this way one obtains the definition of the class of vectors.

In the text, in order not to wander off too much from the usual form of definitions, we have written

$B - A = D - C$

when the relation (4) holds. We have thus *defined* in this way the relation $=$ between two vectors. However, as we have remarked, this is no longer permitted, once one admits Leibniz's universal notion of equality.

For what concerned the forms F_1 (p. 17), the class u consists of all the pairs made up by a point and a real number; the relation α is the following:

$(x_1, A_1; x_2, A_2; \ldots; x_n, A_n)\alpha(y_1, B_1; y_2, B_2; \ldots; y_m, B_m)$

if and only if, for any arbitrary chosen point O, we have

$\Sigma_1^n x_i(A_i - O) = \Sigma_1^m y_i(B_i - O)$

The conditions expressed in (3) are verified etc. One can proceed in the same way for the forms F_1, F_2 (appendix, p. 181).

3

Measuring the size of infinite collections of natural numbers: Was Cantor's theory of infinite number inevitable?

> The possibility that whole and part may have the same number of terms is, it must be confessed, shocking to common sense.
>
> Russell, *Principles of Mathematics*, 1903, p. 358

3.1 Introduction

In this chapter we will not deal directly with abstraction principles but it is nonetheless a pivotal chapter for the connection between abstraction principles and issues related to abstractionist treatments of infinity. For it will be exactly a reflection on alternative ways of measuring the size of infinite sets that will lead us to our considerations on the status of Hume's Principle in Chapter 4.

Two central issues seem to have determined the reflection on mathematical infinity in Western thought. The first concerns its existence. The second whether it can be measured. In this chapter, I will only deal with the second aspect of the issue although, of course, the two issues cannot always be separated. The structure of the chapter is as follows. First, I will retrace some of the major historical positions that were taken with respect to the paradoxical properties displayed by infinite sets of natural numbers with emphasis on whether there could be an arithmetic of infinite sets. In the second part, I will describe recent mathematical developments that offer a way to measure the size of infinite sets of natural numbers while preserving the part–whole principle. In the third part, I will offer some philosophical reflections as to how these recent mathematical developments impact various historical and philosophical claims found in the literature, including Gödel's claim, which I contest, as to the inevitability of Cantor's definition of infinite number. This is the notion of inevitability referred to in the title of this chapter, namely the claim that if one wants to generalize the notion of number from the finite to the infinite there is only one possible way to go and that is the Cantorian notion of cardinal number.

3.2 Paradoxes of the infinite up to the middle ages

While paradoxes of the infinite have often popped up in geometrical contexts—witness the long debates on the one to one correspondence between the points of two segments of different lengths (see Mancosu 1996, chapter 5) or Torricelli's determination that there is a solid of infinite length with finite volume and no center of gravity (see Mancosu and Vailati 1991, Mancosu 1996, chapter 5)—in this chapter I will only focus on the paradoxes concerning the determination of sizes of infinite collections.[1] A simple example will unify this chapter from beginning to end, namely the question of whether the collection of natural numbers and that of the squares of natural numbers (or the even numbers) have the same size. The paradox, epitomized in Galileo's (1939, 1958) *Two New Sciences*, consists in persuading us that

1) There are more natural numbers than squares
2) The collection of natural numbers has as many elements as the collection of squares.

Ad 1: this seems incontrovertibly true, since there are lots of natural numbers that are not squares, 2, 3, 5 and so forth.

Ad 2: this also seems true, as we can arrange the natural numbers in order and write under each one of them the corresponding square, thereby showing that there is a one to one correspondence between the natural numbers and the squares.

We will come back later to the principles that underlie the intuitive pull both in the direction of (1) and (2).

When does the paradox first find its expression? It is actually unclear when the paradox, in the numerical form I just gave, appears. In the Greek tradition we have paradoxes that are related, but are not identical, to it; in this tradition what is claimed to be paradoxical is the existence of different sizes of infinity. Proclus might be the first source we have where such an example is discussed. He proposes a paradox that was raised in connection to the definition of the diameter of a circle:

But if from one diameter two semicircles are produced, and if an indefinite number of diameters can be drawn through the center, it will follow that the number of semicircles is twice infinity. This difficulty is alleged by some persons against the infinite divisibility

---

[1] For good overviews of issues related to the historical developments of the concept of infinity see Levy (1987), Moore (1990), and Zellini (2005).

of magnitudes. We reply that a magnitude is infinitely divisible, but not into an infinite number of parts. The latter statement makes an infinite number actual, the former merely potential; the latter assigns existence to the infinite, the other only genesis. (Proclus 1992, p. 125)

The ultimate assumption, which Proclus seems to subscribe to, is that an infinite cannot be twice as large as another infinite. A similar position is found, for instance, in Philoponus (sixth century A.D.). In *De Aeternitate Mundi*, he touches on the topic in the process of arguing against those who deny that the world has a beginning:

Moreover, suppose the kosmos had no beginning, then the number of individuals down, say, to Socrates will have been infinite. But there will have been added to it the individuals who came into existence between Socrates and the present, so that there will be something greater than infinity, which is impossible. Again, the number of men who have come into existence will be infinite, but the number of horses which have come into existence will also be infinite. You will double the infinity; if you add the number of dogs, you will triple it, and the number will be multiplied as each of the other species is added. This is one of the most impossible things. For it is not possible to be larger than infinity, not to say many times larger. Thus, if these strange consequences must occur, and more besides, as we shall show elsewhere, if the kosmos is uncreated, then it cannot be uncreated or lack a beginning. (Philoponus, *De Aeternitate Mundi Contra Proclus*, ed. Rabe, pp. 9, 11–14, 17; cited in Sorabji 1983, p. 215)

The first occurrence I know of a defense of the existence of different sizes of infinity given in terms of collections of natural numbers comes from the Islamic philosopher and mathematician Thabit ibn Qurra (ninth century A.D.). Ibn Qurra's position comes to us as a report of answers he gave to questions posed by his disciple Abu Musa Isa ibn Usayyid preserved in an Arabic manuscript (at the British Library in London) entitled 'Questions asked of Thabit ibn Qurra al-Harrani' (MS Add. 7473, folio 12b–16b). The text has been recently edited by A. I. Sabra and M. Rashed (Sabra 1997 and Rashed 2009; a previous account was given in Pines 1968):

We questioned him also regarding a proposition put into service by many revered commentators, namely that an infinite cannot be greater than an infinite. He pointed out to us the falsity of this (proposition) also by reference to numbers. For (the totality of) numbers itself is infinite, and the even numbers alone are infinite, and so are the odd numbers, and these two classes are equal, and each is half the totality of numbers. That they are equal is manifest from the fact that in every two consecutive numbers one will be even and the other odd; that the (totality of) numbers is twice each of the two [other classes] is due to their equality and the fact that they (together) exhaust (that totality), leaving out no other division in it, and therefore each of them is half (the totality) of numbers. It is also clear that an infinite is one third of an infinite, or a quarter, or a fifth, or any assumed part of one and the same (totality of) numbers. For the numbers divisible by three are infinite, and they are one third of the totality of numbers; and the numbers divisible by four are one fourth of the totality of numbers; and the numbers divisible by five are one fifth of the totality of numbers; and so on for all other parts of (the totality of numbers). For we find in every three consecutive numbers one that is divisible by three, and in every four consecutive numbers one that is divisible by four, and in every five (consecutive numbers) one that is divisible by five; and in every multitude of consecutive numbers, whatever the multitude's number, one number that has a part named after the multitude's number. (Sabra 1997, pp. 24–25)

This might well be the first occurrence in which an arithmetic of infinite collections comes to the fore. Whatever the complexities related to the interpretation of the text, it is quite obvious that ibn Qurra defends an infinitistic position according to which there are infinite numbers and that an infinite can be larger than another infinite. An intuitive principle deployed by ibn Qurra is that if A and B (disjoint) are such that they together are equal to a collection C then, if the size of A is equal to the size of B, the size of C is twice the size of A (and twice the size of B).

When ibn Qurra states that odd numbers and even numbers have the same size one should be careful not to immediately read his argument as being the standard one based on one to one correspondence, for the motivation adduced does not generalize to other arbitrary infinite sets. Rather, it would seem that some informal notion of frequency (how often do even numbers (respectively odd numbers) show up?) is in the background of ibn Qurra's conception of infinite sizes ("we find in every three consecutive numbers one that is divisible by three").

What would he have replied to the possible objection that there are as many even numbers as natural numbers based on a one to one correspondence between the two collections? The text is silent on these issues.

The Latin middle ages ignored these classical and Islamic sources. As Murdoch puts it, "on the basis of the available evidence, then, philosophers and theologians in the Latin West appear to have realized the importance of the paradox on their own." (Murdoch 1982, p. 569) It is of course here out of the question to rehearse the lively debates found on this topic in the thirteenth and the fourteenth centuries although, with very few exceptions, these authors rarely use natural numbers for their examples.[2] Biard and Celeyrette (2005) provides an extensive selection of

---

[2] In addition to those mentioned in the main text, exceptions include Duns Scotus, Oresme, and Bradwardine. For Scotus's position (*Caveat:* many now think it is not a work by Scotus) see Scotus 1639, Quaestio IX on the third book of the Physics, p. 203. For Oresme see the discussion by Sesiano 1996. The text with the comparison of odd and natural numbers reads: "Omnis multitudo infinita est simpliciter infinita, ergo nulla talis est alia maior vel minor. Consequentia patet, quia non dicitur quod

texts for the fourteenth century (for a recent useful overview and references to the classic literature on the infinitely large (Duhem, Meier, Murdoch, and so forth) see also Dewender 2002; Bianchi 1984, chap. 2, is excellent for the paradoxes of infinity and their context in the thirteenth century).

By and large, one can distinguish four attitudes towards the possibility of unequal actual infinities:

1) Use the paradoxical features that such infinities would have (one is a part of another, i.e. one infinity is smaller than another) to declare this strictly impossible and thus block the process of generation of such infinities (e.g. Bonaventure [Bianchi 1984; Dales 1984]).

2) Accept infinite collections but deny that 'greater than', 'less than' and 'equality' can be applied to infinities (Duns Scotus [Petruzzellis 1966], Oresme

---

unum infinitum simpliciter, id est undique, sit maius e<o>dem consimiliter infinito, et potest argui sicut prius <quod> sibi invicem su<per>posita non excedunt nec exceduntur. Probatur antecedens, et capiatur multitudo numerorum imparium et quadratorum [videtur videtur perfectior]; tunc, cum ibi sit primus, secundus, tertius et quartus et sic sine fine [omnium] <secundum> ordinationem <omnium> numerorum, sequitur quod multitudo numerorum <imparium> non est minor quam multitudo omnium numerorum quorumqumque, et ita argueretur quod multitudo omnium partium proportionalium imparium non est minor tota multitudine parium et imparium, vel tota multitudo partium mediaetatis quam totius" (Kirschner 1997, pp. 260–261). Bradwardine's *Geometria Speculativa* attempts to set up a paradox of infinity using the notion of ratio. In proposition 3.35 he concludes an argument by claiming "therefore all infinities are mutually equal". But the conclusion of proposition 3.36 is that "one infinite may be greater than another". See Bradwardine 1979, pp. 99–100. There is also an attribution of the comparison between odd and natural numbers to Albert of Saxony. As far as I can tell this attribution begins with Maier 1949 (p. 170), who however qualifies the attribution by saying that Albert compares the two sets in a "roundabout" [*umständlicher*] way. The attribution is then repeated without qualification by later authors (including Gericke 1977, p. 54 and Sebestik 1992). The source indicated by Maier is Quaestio X in *Questiones subtilissime in libros Aristotelis de celo et mundo* (Albertus de Saxonia 1492). Once one looks at the text one sees that Maier extrapolated from an example concerning an infinite series of infinitely alternating white and black patches. The thought experiment consists in replacing every black patch by the white one following it: "Arguitur de multitudine. Nam sit unum pedale per imaginationem divisum per partes proportionales. Tunc super primam partem sit aliquod album, super secundam sit aliquod nigrum, et super tertiam iterum sit aliquod album, et super quartam iterum aliquod nigrum, et sic alternatim de aliis partibus proportionalibus. Tunc auferatur primum nigrum, et transferatur secundum album super secundam partem; deinde auferatur tertium nigrum, transferendo sicut prius. Et consequenter amoveantur omnia nigra. Istum casum concederent adversarii. Tunc clarum est quod super quamlibet partem erit aliquod album. Ergo per primam [secundam] suppositionem multitudo alborum non est maior nec minor multitudine partium proportionalium. Et per idem patet ex principio casus quod tota multitudo alborum et nigroum non est maior nec minor multitudine partium proportionalium. Ergo sequitur quod alba et nigra simul sumpta non sunt plura nec pauciora quam alba solum." (quoted according to Sesiano's transcription in Sesiano 1988, pp. 46–47, note 51); compare Sarnowsky 1989, p. 169, note 164. Albert then goes on to generalize to alternations in which a white patch only occurs in position 1, position 1000, position 2000, and so forth, concluding that the white and black patches taken together are not more nor less than the white ones.

[Sesiano 1998; Kirschner 1997], Albert of Saxony [see Gericke 1977; Sesiano 1988; Sarnowsky 1989, pp.149–171]).

3) Accept infinities and analyze how different part–whole relations apply to infinities than those that apply to finite quantities (Henry of Harclay [Murdoch 1981a; Dales 1984], Gregory of Rimini [Cross 1998]).

4) Accept infinities and try to develop an arithmetic of the infinite in analogy to the arithmetic of the finite.

The first text in the Latin West that propounds Option 4, that is, that defends the possibility of comparing different sizes of numerical infinity, is part of the treatise *De Luce* by Robert Grosseteste written about 1220:

Now, it is possible that an infinite collection of number stands to an infinite collection in every numerical ratio and even in every non-numerical ratio. And some infinites are more than others and others less than others; thus, the collection of all numbers, both odd and even, is infinite and it is greater than the collection of all the even numbers, even though this too is infinite, for it exceeds it by the collection of all the even numbers, even though this too is infinite, for it exceeds it by the collection of all the odd numbers. The collection of all numbers proceeding without break from unity by doubling is also infinite, as too is the collection of all the [numbers] corresponding to those doubles as their halves, and yet the collection of the halves must be half of the collection of their doubles. In the same way, the collection of all numbers proceeding without break from unity by tripling is triple the collection of all the thirds corresponding to those triples. The same clearly holds for numerical ratios of every kind, since one infinite can stand to another infinite in any one of these ratios. (Grosseteste, *De Luce*, approximately 1220)[3]

---

[3] The English translation is from Lewis 2013b, p. 241, although I have modified the translation of 'aggregatio' from 'sum' to 'collection'; on Grosseteste see Lewis 2012 and 2013a. The Latin text is edited in Grosseteste (2011) and Panti (2013). It reads as follows: "Est autem possibile ut aggregatio numeri infinita ad aggregationem infinitam in omni numerali se habeat proportione et etiam in omni non numerali. Et sunt infinita aliis infinitis plura et alia aliis pauciora. Aggregatio omnium numerorum tam parium quam imparium est infinita, et ista est maior aggregatione omnium numerorum parium, quae nihilominus est infinita; Excedit namque eam aggregatione omnium imparium. Aggregatio etiam omnium numerorum ab unitate continue duplorum est infinita. Et similiter aggregatio omnium subduplorum illis duplis respondentium est infinita, quorum subduplorum aggregationem necesse est esse subduplam aggregationi duplorum suorum. Similiter aggregatio omnium numerorum ab unitate continue triplorum tripla est aggregationi omnium subtriplorum illis triplis respondentium. Et similiter patet de omnibus speciebus numeralis proportionis, quoniam secundum quamlibet earum proportionari potest infinitum ad infinitum." (Grosseteste 2011, pp. 77–78 and Panti 2013, pp. 228–229) Another passage from Grosseteste, this time from his *Commentary on the Physics of Aristotle*, is relevant here: "Credo tamen quod, sicut alibi diximus, unus numerus infinitus ad alium infinitum numerum se potest habere in omni proportione, numerali et non numerali. Aliquis enim numerus infinitus duplus est ad alium numerum infinitum, et triplus, et sic secundum ceteras species proportionis. Et etiam aliquis numerus infinitus se habet ad alium sicut diameter ad costam, et hoc alibi probatum est. Et iterum audacter dico quod omnis numerus infinitus ipsi Deo (cuius sapientiae non est numerus) finitus est plus quam binarius est mihi finitus. Est illi finitus numerus infinitus collectus ex omnibus paribus, et similiter numerus infinitus collectus ex omnibus imparibus, et similiter omnes numeri infiniti qui infinities possunt diuidi. Sicut enim quae uere in se finita sunt

Grosseteste was followed by Henry of Harclay (see Harclay 2008, Quaestiones XVIII and XXIX) who was however sensitive to the fact that comparison of unequal infinites seemed to contradict the Euclidean axiom that "Omne totum est maius sua parte". Henry declares the axiom inapplicable to infinities and only valid for finite quantities although for Henry the part–whole axiom is "at the same time subordinate to a more general axiom which does apply to infinities" (Murdoch 1981*a*, p. 54). This led the way to Gregory of Rimini's analysis of the different senses of "part", "whole", "greater than" in the context of infinity (on Gregory see Cross 1998).

3.3 Galileo and Leibniz

Galileo and Leibniz reject the possibility of a theory of size of infinite collections, although they do so starting from different assumptions.

As I mentioned at the outset, it was Galileo who, in *Two New Sciences* (1638), gave classic expression to the paradox, using the natural numbers and the squares of natural numbers. Galileo draws the following conclusion from the paradox:

Sagredo: What then must one conclude under these circumstances?
Salviati: So far as I see we can only infer that the totality of all numbers is infinite, that the number of squares is infinite, and that the number of their roots is infinite; neither is the number of squares less than the totality of all the numbers, nor the latter greater than the former; and finally the attributes "equal," "greater," and "less," are not applicable to infinite, but only to finite, quantities. When therefore Simplicio introduces several lines of different lengths and asks me how it is possible that the longer ones do not contain more points than the shorter, I answer him that one line does not contain more or less or just as many points as another, but that each line contains an infinite number. (Galileo 1939, pp. 32–33; 1958, p. 45)

Thus, Galileo's final word corresponded to the positions defended by Duns Scotus, Oresme and others concerning the non-applicability of the relations of equality, less than and greater than, to infinite collections.

nobis sunt infinita, sic quae uere in se sunt infinita illi sunt finita." I report this text following Professor Lewis, who, together with Professor Peter King, is in the process of editing the *Commentaries on the Physics*. (Email communication dated February 19, 2009) Whether there is a significant connection between Thabit ibn Qurra and Grosseteste is unclear. Cecilia Panti who interprets 'aggregatio' as 'sum' thinks the affinity between the two texts is very weak for, according to her commentary to *De Luce* (Grosseteste 2011, p. 110), Thabit ibn Qurra speaks of infinite collections whereas Grosseteste speaks of infinite sums. I am not persuaded by the reading of 'aggregatio' as 'sum' (hence the modification of Lewis' translation) but this issue cannot be resolved here.

Leibniz's position is more radical than that of Galileo in that while admitting that an actual infinite can exist, this is only in the distributive mode and never collectively or as a whole:

There is an actual infinite in the mode of a distributive whole, not of a collective whole. Thus something can be enunciated concerning all numbers, but not collectively. So it can be said that to every even number correspond its odd number, and vice versa; but it cannot be accurately said that the multiplicities of odd and even numbers are equal. (Leibniz, 1875–1890, vol. II, p. 315)

So, Leibniz refuses the existence of an infinite collection taken as a whole to which a size could be attributed. (For more on Galileo and Leibniz on the infinite see Burbage and Chouchan 1993, Knobloch 1999, Leibniz 2001, Arthur 1999, 2001, 2015, Brown 2000, Breger 2008, van Atten 2011, Levey 2015)

3.4 Emmanuel Maignan

A very original position on the nature of the infinite is found in Emmanuel Maignan (1601–1676). He is not a well known figure although he was quite influential on the Spanish enlightenment and left a mark in the history of optics and perspective.[4] Maignan belonged to the order of Minims and taught mathematics, optics and philosophy in Toulouse. In 1648 he wrote a successful *Perspectiva horaria sive de horographia gnomonica tum theoretica tum pratica* that was highly praised and this led to contacts with Mersenne and Fermat. He was decidedly anti-Aristotelian, especially in natural philosophy, and his position on the infinite in fact contradicts Aristotle's denial of the actual infinite.

Maignan treats the infinite in his *Cursus Philosophicus* (1652 [1st edition], 1673 [2nd edition]) and devotes to it a very lengthy dissertation of 30 pages (pp. 283–313 of the second edition; all page numbers below are from the second edition).[5] He is probably the most articulate defender of the existence of actual infinity in the seventeenth century and Cantor took notice of his work (Cantor 1932, p. 405, Note 1; see also the index to Cantor's correspondence with theologians, under 'Maignan', in Tapp 2005). Since he is not very well known, I will devote a little more space to him than to the other authors. Moreover, as his position contains a number of innovative theses that have confused interpreters, I shall begin with the less problematic statements.

---

[4] For his influence on Spanish enlightenment see Israel (2002, pp. 528–531). On Maignan's anamorphosis in the convent of Trinità dei Monti in Rome see Terski (2006) and Pascal (2005).

[5] In the secondary literature, Maignan's treatment of the infinite is only cursorily addressed in Gardies (1984).

Maignan, as I said, defends the existence of infinite collections and the existence of different sizes among infinities. One example of a collection *infinite categorematice* (or *in actu*) given by Maignan is the collection of all possible human beings. The collection of all possible lions can be joined to it showing that an infinite collection can be greater than another. The (possible) eyes of all the possible human beings will be twice as infinite as the collection of all possible human beings. (Maignan 1673, p. 293)

One of his favorite examples of categorematic infinity is the collection of natural numbers. His claim for the existence *in actu* of such a collection consists in pointing out that there is no last finite number and that a multitude that has no last finite number is infinite. (Maignan 1673, p. 285) As pointed out above, he also believes that the relations of greater than, less than and equality hold not only among finite quantities but can also be applied to infinite quantities:

The same thing is confirmed through the previous proposition since, for example, an infinite collection from which units can be subtracted (not only ten but infinitely many) while the collection remains infinite is, obviously, infinitely greater before the subtraction takes place than after; thus, since the collection does not cease being infinite after the subtraction, the infinite will be, as such, infinitely smaller than it was earlier. You could say that this is in conflict with the generally accepted thesis that holds that the terms "greater" and "smaller" can only apply to finite quantities but not to infinite quantities; or, at least, that they can be applied to infinite quantities only in a very improper way. I reply that this idea has the following feature, namely that it is widespread. This fact notwithstanding, I would say, with permission, that it also has this other feature, namely that its ground is nothing else but a false notion of infinity. Moreover, the advantage it offers, which consists in apparently solving some difficulties that are usually put forth by denying that "greater" and "smaller" are properties that can be predicated of infinity, does not subsist for it ends up not resolving the difficulties. (Maignan 1673, pp. 293–294)

The true notion of infinity for multiplicities, according to Maignan is "illud in quo sunt unitates nullo numero finito comprehensibiles" (that in which there are units that cannot be comprehended by any finite number) (Maignan 1673, p. 287).

A further argument propounded by Maignan consists in appealing to Euclid's common notion 9 to the effect that the whole is greater than its part. Taking two collections A and B the collection of the two together is greater than each one taken separately:

In addition, I argue for this thesis in an absolute fashion, for there is a common notion (In Euclid it is the 9th of the first book) which states: the whole is greater than the part; that is, taking as example A and B, the union of the two together is greater than each one of them taken separately and something else that does not belong to it. In effect, with the term 'greater' we understand that which contains everything included in one collection (or the equivalent) with the addition of something else that is not in that collection. With the term 'less' we understand the opposite. Thus, it is evident that in the whole is included all that is in the part (since that is contained in the whole) and, moreover, something else that is not contained in that part, that is, the other part; otherwise, subtracting a single part one would thereby subtract the whole itself, which is evidently false. On the other hand, one can infer from proposition 1, observation 2, that there are infinite parts. And from the words of the previous proposition it results that not only a part of the infinite can be infinite but that two or three and so on until the collection of those infinities become infinite itself, since in every infinite the infinities are infinite. (Maignan 1673, p. 294)

From the above, it is obvious then that for Maignan the collection of even numbers is smaller than the collection of whole numbers, for they stand as part to whole. So far, Maignan presents us with the idea that infinite collections exist and that they can contain infinite parts as sub-collections so that if A is an infinite sub-collection of C, C is greater than A. He obviously thinks that the part–whole principle holds both for finite and for infinite collections. This is not a problem, unless we couple the principle with a form of what is now sometimes called Hume's Principle, according to which two collections have the same size if there is a one to one association between all the elements of the two collections. Trouble then seems to be looming large when, in a corollary, Maignan goes on to give the following account of equality:

From this it follows that two infinites can be equal, just as in an infinite series of pairs there necessarily exist two series of unities completely identical, provided that a unity corresponds always to another unity, so that there is no excess or defect in one or the other series, which corresponds, in turn, to the definition of equality. From here you also grasp that one can properly speak of "quantity" in the sense we have explained but not in that ordinary way in which by "quantity" is usually understood a certain determinate number in its kind corresponding to another in the number of units. (Maignan 1673, p. 294)

On the face of it this risks collapsing all sizes of infinite sets of natural numbers to the same size thus destroying what has been just claimed as a consequence of the part–whole principle.[6] Unless a charitable interpretation of this criterion is

---

[6] Gardies (1984) has in fact read this passage as decreeing that one to one correspondence gives the criterion of equality for infinite sets and has thus concluded to the inner inconsistency of Maignan's thought on this issue. (Gardies 1984, p. 126). Gardies's reading finds some support in an example coming on p. 293 where Maignan plays with a thought experiment of the following sort. Consider all possible human beings. Order them so that they form an infinite sequence. Now consider the same collection of possible human beings but organize them in two distinct rows. Maignan claims that such rows must be infinite. Now, this does not follow but Maignan is probably saying that the rule for forming the two sequences is that we start from the original sequence and we put every second human being in the second sequence. Then Maignan claims that the sequence of pairs of human beings so obtained is infinite but of an infinity smaller than the infinity of possible human beings. I should point out that Gardies seems to misunderstand the text here and claims that Maignan asserts that the pairs are more numerous than the units constituting the pairs.

found, Maignan's criterion of equality applied to natural numbers spells disaster for Maignan's intuition concerning different sizes of infinity. Indeed, consider the following two sequences:

1, 2, 3, 4, 5, . . .
2, 4, 6, 8, 10, . . .

Then, Galileo's paradox is immediately at hand. By the part-whole principle the first collection is larger than the second. By the equality criterion (or Hume's Principle) the two collections are equal in size. We will soon see how Maignan tries to dispel Galileo's paradox and this will show that when pressed he would have given a qualified reading of the equality criterion.

Maignan's discussion of Galileo is introduced abruptly in a section which aims at replying to an objection against the actual infinite. Maignan's description of Galileo's paradox is quite straightforward. However, he interprets the paradox as an argument against his claim that there are infinite collections *in actu*, namely that the collection of natural numbers is a categorematic infinite (or infinite *in actu*; Maignan seems to use the two expressions interchangeably). Obviously, this was not Galileo's take on the matter and Maignan is clear about this. Indeed, Galileo accepts the existence of different infinite collections (numbers, squares, non-squares) but claims that the relations 'greater than', 'smaller than', and 'equal' cannot be applied to infinite collections.

Maignan begins his discussion by first rehearsing Galileo's considerations on the frequency of squares and non-squares leading to the claim that there are more non-squares than squares (let us call this 'Claim 1').[7] But the heart of the paradox begins in what follows and for simplicity I will label the claims in the argument.

It is evident that the collection of all numbers (containing both squares and non-squares) is greater than the collection of squares since the whole is greater than its part (Claim *).[8]

Furthermore, "it is also evident that the collection of all squares is equal to the collection of all roots, since each root is the root of a square and each square has only one root". (Claim 2).[9] Note that there is a certain ambiguity in Maignan's text that is not in Galileo. While Galileo says "loro [i quadrati] esser tanti quanto le loro radici", Maignan speaks of equality [aequalem esse] but this is ambiguous between having the same size and being identical as collections.[10]

---

[7] Compare Galileo 1939, p. 32.

[8] Compare Galileo 1939, p. 32.

[9] Compare Galileo 1939, p. 32.

[10] Galileo says: "If I should ask further how many squares there are one might reply truly that there are as many as the corresponding number of roots, since every square has its own root and every root its own square, while no square has more than one root and no root more than one square." (Galileo 1958, p. 44; compare Galileo 1939, p. 32).

With those premises in place a contradiction is now at hand:

But these two things [Claims * and 2] are contradictory since the collection of all the roots is equal to the collection of all the numbers [Claim 3],[11] for both the squares and the non-squares are roots, as they can be multiplied by themselves. Consequently, the collection of the squares is equal to the collection that includes at the same time the non-squares and the squares, which is contrary to what has been formerly established [Claim %].[12] (Maignan 1673, p. 304)

More formally, letting A be the collection of all roots, B the collection of all numbers, and C the collection of squares, Maignan sees the argument as proceeding through the following steps:

$C = A$ [Claim 2]
$B = A$ [Claim 3].

Thus,

$C = B$ [Claim %; according to the principle that those which are equal to a third are also equal to one another using Claim 2 and Claim 3; obviously the equality sign here is used as a short hand for whatever Maignan means by equality.]

But this means that Part C is equal to the Whole B, contrary to what has been established in Claim *. Contradiction.

Notice that in Maignan's reconstruction of the paradox Claim 1 is not playing any role, just as it does not in Galileo's account where it is used only to strengthen the sense of amazement occasioned by the main paradox. Nonetheless, it will be useful to look at Maignan's reply to Claim 1.

Maignan tries to dismantle the Galilean arguments by rejecting 1) that the non-squares are more than the squares; and 2) that the roots are equal to the numbers (namely, Claim 3).

Ad 1) Maignan provides an alternative way of counting the squares. He first orders the non-squares into a sequence:

3, 5, 6, 7, 8, and so forth.[13]

Then he creates the following grid obtained by squaring successively all the numbers in each column:

---

[11] Once again Galileo is clearer in speaking not of equality but more clearly says "non si può negare che elle [le radici] siano quante tutti i numeri." (Galileo 1958, p. 44; Compare Galileo 1939, p. 32).

[12] Galileo says "converrà dire che i numeri quadrati siano quanti tutti i numeri." (Galileo 1958, p. 44; compare Galileo 1939, p. 32).

[13] For some reason I cannot figure out, Maignan thinks of 2 as a square and does so on two different occasions; however, when computing the number of squares between 1 and 10 or between 1 and 100 he gives the right answer.

3	5	6	7	8	...
9	25	36	49	64	...
81	625	1296	...
...	...	...

He then notices that under every nonsquare there are infinitely many squares and concludes that there are more squares than nonsquares:

> Therefore, the nonsquare numbers are far from being more numerous than the squares; on the contrary, for each nonsquare there are infinitely many squares. (Maignan 1673, p. 305)

Ad 2) Maignan proceeds by distinguishing two meanings to the claim that "the collection of roots is equal to that of all numbers". In the first sense one can consider all the numbers taken *collective* (i.e, as a whole) and in the other *distributive*. Maignan argues, confusingly, that in the first case there is an evident contradiction, for reading 2 with the *collective* reading would lead to the conclusion that there are numbers beyond the numbers.[14] More interesting is the objection to the second reading, namely the *distributive* reading:

> In the second meaning, one does not prove that the collection of squares (although to each root corresponds a square and vice versa) is equal to the collection of all the numbers. One does not prove it, I say, unless in addition to supposing in this comparison that every square is a root, one also assumes, conversely, that every root is a square. But this would be admitted without justification, since it is evidently false; in effect, there are many numbers that are both non-squares and roots. Thus, for this reason, although one could claim,

---

[14] I must admit that the argument puzzles me. Here is the text: "Deinde ad 2 obiectionis partem distinguo id quod ibi simpliciter astruitur, multitudinem scilicet radicum aequalem esse multitudini numerorum omnium; cum enim haec multitudinem aequalitas fundetur in hoc quod quilibet numerus sit in seipsum multiplicabilis, adeoque sit radix; ea propositio potest habere hunc duplicem sensum. Primus est, omnes numeri collective sumpti sunt radices, ita ut omnes ac solae radices sint omnes numeri collective sumendo omnes, nulloque praetermisso numero. Secundus est; omnes numeri distributive sumpti sunt radices; ita ut nullus in particulari sit numerus, qui non sit radix alicuius alterius: primus sensus manifestam habet contradictionem; quia sic praeter omnes numeros essent aliqui numeri; cum enim numeri non possint sibi invicem esse radices velut circulari regressu; sed alii aliorum semper ulteriorum sint radices; necesse est ut dum sumuntur omnes sed solae radices, supponantur omnes earum omnium producti: at quia nulla est, ut dixi, suus productus; necessario praeter solas radices est aliquis productus: Ergo si nihilominus omnes ac solae radices sunt omnes numeri, hoc est si dicendo eas, dicis omnes numeros; praeter omnes numeros dicis unum numerum: ubi est adhuc altera contradictio; quia ille unus ulterior, non solum erit numerus praeter omnes numeros; sed etiam erit radix praeter omnes radices, nempe erit radix alterius numeri ulterioris. Ergo hoc primo sensu dici nullatenus potest multitudinem radicum omnium aequalem esse multitudini omnium numerorum." (Maignan 1652, vol. III, pp. 1042–43; 1673, vol. III, p. 305) The same solution to the paradox is found in Maignan's disciple Jean Saguens who also concluded that from the fact that two different progressions make the number of squares greater, respectively smaller, than the non-squares only follows that "we do not understand an infinite multitude but it does not follow that it is impossible." (Saguens 1707, vol. III, disp. X, art. VI, p. 122).

according to the other point of view (i.e. according with the first meaning that I already proved false), of all the roots but only of them that they are all the numbers, nonetheless one cannot claim that the roots are all the square numbers. Now, since even so the roots are numbers one cannot state that the numbers that are strictly squares are as many as all numbers. (Maignan 1673, p. 305)

It seems to me that Maignan's argument here shows that one cannot read the criterion of identity we encountered earlier unqualifiedly. His notion of equality for infinite collections is stronger than mere one to one correspondence since in the argument we have just given he denies that squares and roots are equal in size (let alone extensionally equal) despite the one to one correspondence between the two collections. Indeed, judging from this objection he seems to be forced to accept a criterion of equality between collections that might simply turn out to be extensional equality.

It should be obvious by now that Maignan's position is quite unstable and that Maignan did risk to "wander without end" and to shipwreck in "an immense and dangerous sea", to use two beautiful expressions from his preface. In his reply to (1), he exploits the reordering of the sequence of natural numbers to argue that for every non-square there are infinitely many squares but fails to penetrate the logic of such reorderings with infinite sets which allows one to give apparently conclusive arguments for two contradictory claims (there are more squares than non-squares and vice versa). Indeed, his answer to (1) left him hostage to the possible retort that he had not really explained why on the Galilean counting there are more non-squares than squares and how this fact can be reconciled with his calculations. He addresses the topic by claiming that his calculation is no less precise than the Galilean one and concludes:

Now, how this can be the case is known only to Him who knows perfectly the nature of infinity. We will be satisfied knowing that this is in fact the case. (Maignan 1673, p. 306)

In his answer to (2) he is obviously relying on the notion of part–whole as the appropriate one for defining a measure of size on infinite collections. According to that logic 'equality' between squares and numbers could only hold if every number is a square; one to one correspondence is not enough. But, if so, then his approach is severely limited and fails when one needs to compare collections that do not stand in an inclusion relation such as the collection of even numbers and the collection of odd numbers.

Notwithstanding all the problems in Maignan's conception of the infinite there is no doubt as to his commitments to infinite collections and to the idea that infinite collections of natural numbers can be measured and come in different sizes.

3.5 Bolzano and Cantor

If Galileo and Leibniz seem to agree, *pace* Maignan, that one should not attempt to develop a theory of sizes of infinity, with Bolzano and Cantor this possibility is admitted.

However, Bolzano and Cantor differ on the criteria that determine sizes of infinite collections.

We find an interesting take on different sizes of infinity in Bolzano's *Wissenschaftslehre* (Bolzano 1837, 1973, 2014). Bolzano offers an example constructed by a nested sequence of infinite countable sets and states that each one of the sets in the sequence is infinitely smaller than the preceding one. The example is provided in the context of an argument aimed at showing that infinite sizes are needed to measure the extension of ideas. In order to show that there are ideas with infinitely many objects falling under them Bolzano gives his nested sequence:

It can be shown that there are infinitely many ideas which are of such a nature that one is surpassed in width [*Weite*] by the next. From this it follows that the unit that serves to measure one cannot be used to measure another, so that no finite set of units of measurements suffices to measure the width of all ideas. It seems to me that the following example proves the truth of this claim: Let us abbreviate the concept of any arbitrary integer by the letter n. Then the numbers $n, n^2, n^4, n^8, n^{16}, n^{32}, \ldots$ express concepts, each of which includes infinitely many objects (namely, infinitely many numbers). Furthermore, it is clear that any object that stands under one of the concepts following n, e.g. n^{16}, also stands under the predecessor of the concept, n^8. It is also clear that very many objects that stand under the preceding (n^8) do not stand under the following (n^{16}). Thus of the concepts $n, n^2, n^4, n^8,$ $n^{16}, n^{32}, \ldots$ each is subordinated to its predecessor. It is, furthermore, undeniable that the width of any of these concepts is *infinitely larger* than the width of the concept immediately following it. (And this holds even more for concepts that follow later in the series.) For, if we assume that the largest of all numbers to which we want to extend our computation is N, then the largest number that can be represented by the concept n^{16} is N and thus the number of objects that it includes is equal or smaller than $N^{1/16}$ and likewise the number of objects that stand under the concept n^8 is equal or smaller than $N^{1/8}$. Hence the relation between the width of the concept n^8 and that of the concept n^{16} is $N^{1/8} : N^{1/16} = N^{1/16} : 1$. Since $N^{1/16}$ can become larger than any given quantity, if N is large enough, and since we can take N as large as we please, and since we come closer and closer to the true relation between the widths of the concepts n^8 and n^{16}, the larger we take N, it follows that the width of the concept n^8 surpasses infinitely that of the concept n^{16}. Since the series $n, n^2,$ $n^4, n^8, n^{16}, n^{32}, \ldots$ can be continued indefinitely, this series itself gives us an example of an infinite series of concepts, each of which is of infinitely greater width than the following. (Bolzano 2014, §102)

In *Paradoxes of the Infinite* (1851, 1975*a*), Bolzano recognizes that infinite sets have the characteristic property of standing in one to one correspondence with a proper subset of themselves. However, he refuses to deploy this property for capturing the idea of size of infinite sets:

Therefore, merely for the reason that two multitudes A and B stand in such a relation to one another that to every part *a* occurring in one of them A, we can seek out according to a certain rule, a part *b* occurring in B, with the result that all the pairs (*a+b*) which we form in this way contain everything which occurs in A or B and contain each thing only once—merely from this circumstance we can—as we see—in no way conclude *that these two multitudes are equal to one another if they are infinite* with respect to the plurality of their parts (i.e. if we disregard all differences between them). (Bolzano 1851, section 21; translation in Russ 2005, p. 617)[15]

That even the great Bolzano could have been so far from Cantor led Jan Berg in the introductory notes to Bolzano (1973, pp. 26–28) to rescue the good reputation of Bolzano by pointing out that "in a letter to his pupil Robert Zimmermann on March 9, 1848, that is a few months before his own death" Bolzano gave up his former position. Let me quote the relevant passage from the letter to Zimmermann in full:

Wissenschaftslehre, vol. I, p. 473. The thing has not only become unclear but, as I have just come to realize, completely false. If one designates by *n* the concept of an arbitrary whole number, or better said, should one represent by means of the sign *n* any arbitrary whole number, then it is therewith already decided which (infinite) sets of objects the sign represents. This fact is not at all affected by our request that through the addition of an exponent as n^2, n^4, n^8, n^{16}, ... each number should now be raised to the second power, now the fourth power, and so on. The set of objects represented by *n* is always exactly the same as before although the objects themselves represented by n^2 are not quite the same as those represented by *n*. The false result was due to the unjustified inference from a finite set of numbers, namely those not exceeding the number N, to *all* of them. (Bolzano, 1978, pp. 187–188; my translation; also partially quoted by Berg in Bolzano 1973, p. 27)

Berg goes on to conclude:

Hence, it seems that in the end Bolzano confined the doctrine that the whole is greater than its parts to the finite case and accepted isomorphism as a sufficient condition for the identity of powers of infinite sets. This is a second achievement of major importance in Bolzano's investigation of the infinite. (Bolzano 1973, pp. 27–28; compare 1978, p. 188, note 451)

Thus, Bolzano saved his mathematical soul in extremis and joined the rank of the blessed Cantorians by repudiating his previous sins. While this could be argued

---

[15] See also Bunn (1977), Spalt (1990), and Parker (2009) (section 3). A full analysis of Bolzano's position should also take into consideration the *Grössenlehre* (see Bolzano 1975*b*). See Sebestik (2002) for the best encompassing treatment of Bolzano's philosophy of logic and mathematics.

for sets of natural numbers (see Bolzano 1851, section 33, where Bolzano seems to contemplate using one to one associations as determining the size of certain sequences, such as the sequence of numbers and its squares), the claim strikes me as implausible, if not downright false, when it comes to Bolzano's handling of infinite sets in geometrical contexts.

Without wanting to pick on Berg, I must observe that the literature on infinity is replete with such 'Whig' history.[16] An author is praised or blamed depending on whether or not he might have anticipated Cantor and naturally this leads to a completely anachronistic reading of many of the medieval and later contributions (this was certainly the case with Duhem's interpretation of Gregory of Rimini and Maier's interpretation of Albert of Saxony; recent scholarship has been more cautious (Murdoch, Dewender etc.)).

As we know, it was Dedekind who exploited the property of reflexivity of infinite sets, namely that they can be put in one to one correspondence with a proper subset of themselves, and turned it into a definition of infinity. But it was left to Cantor to use the criterion of one to one correspondence to analyze the notion of size for infinite sets.[17] When faced with the traditional paradoxes of infinity, Cantor drops the intuition that if A is properly included in B then the size of A must be strictly less than the size of B. In the *Mitteilungen* (Cantor 1932, p. 417) this problem comes up in a revealing form. Cantor says:

Let M be the totality (n) of all finite numbers n, M' the totality $(2n)$ of all even numbers $2n$. Here it is definitely correct to say that according to its entities M is richer than M'; indeed, M contains in addition to the even numbers, which make up M, also all the uneven numbers M''. On the other hand, it is also definitely correct that both sets M and M', according to sections 2 and 3, have the same cardinal number. Both (propositions) are certain and they do not conflict with each other if one carefully observes the distinction between reality and number. One should therefore say: the set M has more reality than M', because it contains as parts M' and M'' in addition; the cardinal numbers corresponding to them are however equal. When will these easy and enlightening truths be finally acknowledged by all thinkers? (Cantor, *Mitteilungen zur Lehre vom Transfiniten*, 1887–1888, p. 417 of the 1932 edition; cf. Tapp 2005)

Unwittingly, Cantor shows in this passage how his solution to the Galilean paradox leaves one of our intuitions about infinite sets without proper explication. According to Cantor's theory, all infinite subsets of the natural numbers have the same cardinality, we would say 'size'. But when he tells us that there is a sense

---

[16] In the case of Bolzano this has been rightly emphasized by Spalt (1990, pp. 199–200).

[17] I will assume familiarity with Cantor's theory. For scholarly accounts see Hallett (1984), Purkert (1987), Dauben (1990), Ferreirós (2007). In connection to the topic of this Chapter see also the careful historical analysis in Parker (2009, section 4).

in which the set of natural numbers has 'more reality' (or 'is richer') than the set of even numbers, we feel the pull of our original intuition again: can we put a measure, a 'size', on how much 'more reality' the natural numbers have in comparison to the even numbers? Or the even numbers in comparison to the multiples of 4? While Cantor is clear that there are two notions at play here, he does not indicate that one of the two notions (the one related to 'more reality') can be developed further to allow for a quantitative measurement of the notion of 'more reality'.

Notice that Cantor quite correctly claims that there is no conflict between the two notions of 'having more reality' and 'having the same cardinal number'. My sense is that on account of the fruitfulness of the Cantorian approach in set theory and the lack of interesting alternatives, the general conviction is simply that an interesting theory that generalizes the notion of 'having more reality' to a full blown arithmetic of infinite sets satisfying the part–whole principle cannot be had. I now want to show that recent mathematical work gives us theories that can (*cum grano salis*) be seen as formally capturing (parts of) the intuitive concept of infinity found in Thabit ibn Qurra, Grosseteste, Maignan, Bolzano and all those who believed that the size of the natural numbers is larger than the size of the even numbers.

3.6 Contemporary mathematical approaches to measuring the size of countably infinite sets

Before explaining the recent developments I have in mind, I would like to address and dispose of some reasonable questions that might have been occasioned by the previous discussion. If all one is calling for is an account of the fact that numbers divisible by 2 are more numerous than numbers divisible by 3, and similar examples, then two options seem readily available.

The first option concerns the possibility of using the notion of asymptotic density, as used in number theory,[18] as a mathematical tool for discriminating sizes of infinite sets of natural numbers. If A is any set of natural numbers, let $c_A(n)$ denote the number of objects in A restricted to $[1 \ldots n]$. Thus $c_A(n)/n$ represents the fraction of the first n natural numbers that are in A. If $c_A(n)/n$ approaches a limit, d, as n approaches infinity, then d is said to be the asymptotic density of A. According to this notion the set of even numbers has density $1/2$, the set of odd numbers has density $1/2$, and that of numbers divisible by 3 has density $1/3$. While such an approach helps account for some of the intuitions we encountered

---

[18] Among the many texts in this area see Fine and Rosenberger (2007).

in our historical survey, one of its limitations is that it does not give us a notion that generalizes relations of size among finite sets. Indeed, all finite sets in this approach have the same asymptotic density, namely 0. Some infinite sets have no asymptotic density (as their lim sup and lim inf do not coincide), and many infinite sets (such as the primes) have density 0. Moreover the part–whole principle is not respected as the set of even numbers and the set containing 1 and the even numbers have the same density. While the latter is per se not a criticism of asymptotic density, our aim is in fact to see whether the part–whole principle can be implemented. Even with this goal in mind, asymptotic density can help in providing intuitive constraints on models of theories satisfying the part–whole principle for arbitrary sets. This is what happens in one of the mathematical accounts to be discussed next (see section 6.1).

The second suggestion could be to appeal to standard theorems about binary relations and claim that since $\subset$ is a partial order on the power set of the natural numbers we can extend it to a total order, perhaps satisfying additional properties (see Szpilrajn 1930 and, among the many extensions, Dushnik and Miller 1941, Duggan 1999). This would provide us with a theory that has a $<$ relation (a linear ordering) extending $\subset$ for which trichotomy holds. This is true but completely uninteresting for we have no guarantee as to what relations of size between sets of natural numbers (both finite and infinite) the relation $<$ will induce. It will preserve intuition for sets that stand in the inclusion relation but it will be completely arbitrary on sets that are disjoint or that only partially intersect. Parker (2009) also mentions the great limitations of this approach as a plausible account of size for infinite sets.

Obviously, from an 'arithmetic' theory of sizes of arbitrary sets of natural numbers we would at least expect the theory not to induce results about the size of finite sets that are in conflict with previously established results on the cardinality of finite sets. Moreover, we would also like the preservation of certain basic intuitions concerning the relative size of finite and infinite collections, although how much one ought to expect is of course open for debate.

The approaches to be discussed next are motivated by the aim of preserving the part–whole intuition in ways that also preserve other algebraic properties of finite cardinalities. I claim no originality in my exposition of these theories, indeed in some cases I follow the definitions and the original exposition verbatim. All I hope to do is to convey clearly to the reader the main ideas of such approaches.

3.6.1 Katz's "Sets and their Sizes" (1981)

An interesting mathematical approach to our problem has been explored in a dissertation written in 1981 at M.I.T. by Fred M. Katz, "Sets and their Sizes". Let me quote from its abstract:

Cantor's theory of cardinality violates common sense. It says, for example, that all infinite sets of integers are the same size. This thesis criticizes the arguments for Cantor's theory and presents an alternative. The alternative is based on a general theory, CS (for Class Size). CS consists of all sentences in the first-order language with a subset predicate and a less-than predicate which are true in all interpretations of that language whose domain is a finite power set. Thus, CS says that less than is a linear ordering with highest and lowest members and that every set is larger than any of its proper subsets. Because the language of CS is so restricted, CS will have infinite interpretations. In particular, the notion of one–one correspondence cannot be expressed in this language, so Cantor's definition of similarity will not be in CS, even though it is true for all finite sets. We show that CS is decidable but not finitely axiomatizable by characterizing the complete extensions of CS. CS has finite completions, which are true only in finite models and infinite completions, which are true only in infinite models. An infinite completion is determined by a set of remainder principles, which say, for each natural number, n, how many atoms remain when the universe is partitioned into n disjoint subsets of the same size. We show that any infinite completion of CS has a model over the power set of the natural numbers which satisfies an additional axiom: OUTPACING. If initial segments of A eventually become smaller than the corresponding initial segments of B, then A is smaller than B. Models which satisfy OUTPACING seem to accord with common intuitions about set size. In particular, they agree with the ordering suggested by the notion of asymptotic density.

Katz's starting point is the conflict between two principles of size. The first is ONE–ONE and the second is SUBSET.

ONE–ONE: Two sets are the same size just in case there is a one-one correspondence between them (Katz 1981, p. 1).

SUBSET: If one set properly includes another, then the first is larger than the second (Katz 1981, p. 2).

Conflict occurs when both are used to capture the same notion of size. Since ONE–ONE has won against SUBSET on account of Cantor's successful theory of infinite sets, Katz sets out to develop an approach that will vindicate SUBSET even for infinite collections and in particular for the collection of subsets of natural numbers. His dissertation is complex and full of interesting things but here I have to restrict myself to giving a bare outline.

The starting idea is to define a theory CS (class size) that will contain SUBSET (but in which ONE–ONE is not expressible) and in which the ordering relation between sets is defined in ways that mirror plausible principles of size. The theory's language is a first-order language $L_{CS} = \{\emptyset, I, \text{Atom}(x), \text{Unit}(x), \subset, -, \cup, \cap, <\}$.

The theory CS contains all the axioms for an atomic Boolean algebra and other principles for size, such as the formal version of subset: if $A \subset B$ then $A < B$. Indeed, CS contains much more and it is first characterized model-theoretically as the set of sentences true in all standard finite interpretations of L_{CS}. The latter is defined as follows.

A is a *standard interpretation* of L_{CS} iff

(i) $dom(A) = P(x)$ for some x (x is said to be the basis of the interpretation A; $P(x)$ is the power set of x)
(ii) $A(I) = x$; $A(\emptyset) = \emptyset$; $A(a \subset b) =$ true iff $a \subset b$.

A is a *standard finite interpretation* of L_{CS} iff in addition to being a standard interpretation, it also satisfies that *A* has a finite basis and the following three conditions:

$A(a < b) =$ true iff $card(a) < card(b)$
$A(a \sim b) =$ true iff $card(a) = card(b)$ (where $\sim$ is a defined notion)
$A(Unit(a)) =$ true iff $card(a) = 1$.

The theory CS is defined model-theoretically as the class of sentences that are true in all standard finite interpretations of L_{CS}. A great part of Katz's work is devoted to effectively axiomatizing CS and he succeeds by adding certain division principles to an intuitive theory that includes all the axioms for an atomic Boolean algebra and various size principles. The proof is far from trivial. Sentences that are in CS include, in addition to the already cited SUBSET, other principles such as trichotomy ($x < y$ or $x \sim y$ or $x > y$). Moreover the predicate Sum allows one to state basic principles of size-addition for disjoint sets.

The question as to whether there are infinite models of such a theory receives a positive answer as the theory has models of arbitrarily large finite cardinality and any such first-order theory must have an infinite model. Moreover, Katz shows that there are completions of CS (which is a decidable but incomplete theory) with infinite models over the natural numbers that satisfy a principle he calls OUTPACING. Intuitively the principle says, for two sets x and y, that if there exists an n such that for all $m > n$, the restriction of x to m ($x[m]$) is greater in cardinality than the restriction of y to m ($y[m]$) then $x > y$. Talking about cardinality of such restrictions is unproblematic as all restrictions involved are finite. The models satisfying OUTPACING are more satisfactory than other infinite standard models for they avoid pathologies such as decreeing that there are fewer even numbers than prime numbers and in addition match other appealing intuitive facts related to asymptotic density. Thus, in such models of CS we have pleasing results such as the fact that the size of the even numbers is greater than the size of the numbers divisible by 3, and that the set of squares is smaller than the set of multiples of k, for any $k > 0$. Moreover, every two infinite sets of numbers are such that they are comparable in size on account of trichotomy. However, certain determinations of sizes, such as whether the odd numbers and the even numbers have exactly the same size, depend on the choice of ultrafilter used for coming up with the models of CS + OUTPACING.

In CS we have a defined notion *sum* (x, y, z) that obeys the obvious principles for disjoint sets. But notice that the arithmetical laws valid in Katz's models are different from the standard ones. For instance, we have a maximum element I (a "largest infinite") and thus the arithmetical operations in such models do not preserve all the standard arithmetical rules. By contrast, the theory to be investigated next offers a generalization of all the ordinary arithmetical rules also to infinite sets of numbers.

Katz's thesis has remained little known but it certainly has relations, unbeknownst to the respective authors, with the approach to be discussed next. These relations deserve to be investigated but this is not something I can do in this chapter.

3.6.2 A theory of numerosities

Let us again articulate the principles of size that hold for finite sets. We could call the first the PW principle (part-whole principle):

PW: If A is a subcollection of B then $s(A) < s(B)$
(This corresponds to Katz's SUBSET).

We then have Cantor's principle:

CP: $s(A) = s(B)$ iff there is a one to one correspondence between A and B
(This corresponds to Katz's ONE–ONE).

Notice that everyone accepts these principles for finite sets. The problem only emerges when we try to extend these principles to infinite sets. The paradox that has haunted the history of the infinite can be captured immediately as follows. Assume the principles PW and CP hold for infinite sets. Let B be the set of natural numbers. Let A be the set of even numbers. Since $A \subset B$ by PW we have $s(A) < s(B)$. But A and B can be put in one to one correspondence. So, by CP, $s(A) = s(B)$. Hence $s(A) < s(A)$. Contradiction.

As is well known, Cantor gives up PW for infinite sets and holds on to CP. What happens to PW principle in Cantor's theory? It is weakened to:

WPW: $A \subseteq B$ implies $s(A) \leq s(B)$.

The theory to be described now, which extends PW to all countable sets, originated with Benci (1995), who then joined ranks with Di Nasso and Forti in extending the scope of the approach also beyond the countable sets. The three key papers are Benci and Di Nasso (2003); Benci *et al.* (2006); and Benci *et al.* (2007) (see also Di Nasso and Forti 2010). A useful informal exposition is also found in Gilbert and Rouche (1996).

The progression of the three articles corresponds to an extension from countable sets, to arbitrary sets of ordinals, and finally to "universes", that is, superstructures $V(X) = \bigcup_{n \in N} V_n(X)$ over a base X of size less than $\aleph_\omega$. I will only emphasize the approach for countable sets. Informally the approach consists in finding a measure of size for countable sets (including thus all subsets of the natural numbers) that satisfies PW. The new 'numbers' will be called 'numerosities' and will satisfy some intuitive principles such as the following: the numerosity of the union of two disjoint sets is equal to the sum of their numerosities.

Let us begin by stating what goes on when we assign a size function to a collection of sets. Basically we can think of such assignment as being a triple $< S, (\mathcal{N}, \leq), \nu >$ where:

S is the family of sets whose 'numerosity' we want to count;
$(\mathcal{N}, \leq)$ is a linearly ordered set of numbers (on which addition and multiplication are defined);
ν is a function from S onto $\mathcal{N}$.

Here are some properties one would like the system to have:

(1) if there is a 'bijection' between A and B then $\nu(A) = \nu(B)$
(2) if $A \subset B$ then $\nu(A) < \nu(B)$
(3) If $\nu(A) = \nu(A')$ and $\nu(B) = \nu(B')$ then the corresponding disjoint unions (∇) and Cartesian products ($\times$) satisfy:

$$\nu(A \nabla B) = \nu(A' \nabla B'); \ \nu(A \times B) = \nu(A' \times B')$$

While $< $ Fin, $(N, \leq), \# >$—with 'Fin' denoting the finite sets, 'N' denoting (also in the rest of this chapter) the natural numbers, and '$\#$' denoting the ordinary cardinality on finite sets—satisfies all three properties, this is not the case for the following two examples concerning the class of all sets and the class of all well ordered sets:

a) $< $ Sets, $(\text{Card}, \leq), \|\ \|>$ [$\| A \|$ is the unique cardinal equipotent to A]
b) $< $ WO, $(\text{Ord}, \leq), |\ | >$ [$|A|$ is the order type of A]

The problem is: can we find a system of counting countable sets that satisfies (1), (2), and (3)? The answer is yes but first we need to explain the informal idea behind it.

COUNTING

Suppose we are playing 'tombola', which is the Italian name for what in the United States is called 'bingo'. Tombola is played with 90 wooden pegs numbered 1–90 that are extracted from a bag (one at a time) and placed on a master counter. Suppose I want to check that in the excitement of the game I have not lost one or more of the pegs.

Here are three possible ways of counting:

(a) I place the pegs in the master counter in any order. If there is a one–one correspondence between the pegs and the places in the master counter (which are also numbered), I have shown equinumerosity. This corresponds to Cantor's notion of cardinality.

(b) I list the pegs in the master counter in their 'natural ordering'. Here the peg with the number 17 will be placed in the place numbered 17 in the master counter, and so forth. This reproduces the ordering and it is basically Cantor's ordinal approach.

(c) I list the pegs as follows: on the place numbered 10 in the master counter I put all pegs (one on top of each other) from 1 to 10; in the place numbered 20 all pegs from 11 to 20 and so on until on the place numbered 90 on the master counter I put all pegs from 81 to 90. I can then count the nine piles, each one containing 10 pegs, and the sum yields 90.

It is the intuition behind this third way of counting that is at the source of the new strategy for counting infinite sets. First of all notice that the strategy in c) is independent of the fact that it is the numbers from 1 to 10 that are put in square 10. Had I placed ten randomly picked pegs on 10, 10 more randomly picked pegs on 20, and so forth, the result would have been the same. This leads to three sequences:

1, 2, 3, 4, 5, 6, 7, 8, 9, 10, 11, 12, 13, 14, 15, 16, 17, 18, 19, 20, ..., 30, ..., 80, ..., 90
0, 0, 0, 0, ... 10, 0, ... 10, ..., 10, ..., 10, ..., 10
0, 0, 0, 0, ... 10, 10, 10, 10, 10, 10, 10, 10, 10, 10, 20, ..., 30, ..., 80, ..., 90

The second sequence indicates that in 'box' number 10 we have added 10 units; 10 more in 'box' 20, and so forth. The third sequence gives the partial sums; one holds on to the previous sum until a box n in which new pegs are added is reached and then one adds the number of the new pegs to the previous partial sum. Thus, the basic idea is to split a set of objects into boxes each one containing only finite many objects. The metaphor of putting things in box number 10, 20, and so forth, will be captured by the idea of a labeled set. From now on we deal only with countable sets.

Definition 6.1.   *A labeled set is a pair $< A, l_A >$ where the domain A is a (countable) set and the labeling function $l_A : A \to N$ is finite to one. What that means is that there is a labeling of the elements of A such that only finitely many elements of A can be mapped to the same natural number n.*

Thus A can be re-obtained as the union of the following chain:

$$A_0, A_1, A_2, \ldots A_n, A_{n+1}, \ldots$$

where $A_n = \{a : l_A(a) \leq n\}$.

Considering the finite cardinality of each A_n, that is, $\#A_n$, we can think of $\#A_n$ as the n-th approximation to the "numerosity" of A. The sequence $\gamma_A : n \to \#A_n$ is called the approximating sequence to the numerosity of A.

Obviously this way of counting depends on how a set is labeled although in the finite case the labeling makes no real difference. It is when we move to infinity that change of labeling becomes problematic. For this reason, and not to get confused from the outset, we will now move to counting sets of natural numbers using the 'canonical labeling' $l(n) = n$.

Example: consider the set of even numbers with the labeling function being the identity function: $< \text{Even}, l_{Even}(x) >$.

Then:

$\text{Even}_0 = \{a : l_{Even}(a) \leq 0\} = \{0\}$
$\text{Even}_1 = \{a : l_{Even}(a) \leq 1\} = \{0\}$
$\text{Even}_2 = \{a : l_{Even}(a) \leq 2\} = \{0, 2\}$

$\ldots$

and $\#\text{Even}_0, \#\text{Even}_1, \#\text{Even}_2, \ldots$ that is, 1, 1, 2, 2, and so forth, gives the approximating sequence to the numerosity of $< \text{Even}, l_{Even}(x) >$.

We can now define various relations between labeled sets.

$\mathbf{A} = < A, l_A >$ is a labeled subset of $\mathbf{B} = < B, l_B >$, written $\mathbf{A} \subseteq \mathbf{B}$, iff $A \subseteq B$ and $l_A(a) = l_B(a)$ for all a in A. Similarly for strict inclusion $\mathbf{A} \subset \mathbf{B}$.

Definition 6.2.   *Isomorphism of two labeled sets:*
Two labeled sets $\mathbf{A} = < A, l_A >$ and $\mathbf{B} = < B, l_B >$ are isomorphic iff there exists a bijection $\Phi : A \to B$ that preserves the labeling, i.e. such that $l_B \circ \Phi = l_A$.

Notice that there are lots of non-isomorphic finite labeled sets of the same cardinality. $< \{a\}, l_A >$ and $< \{b\}, l_B >$ are isomorphic just in case $l_A(x) = l_B(x)$.

We can now define the sum and product of two labeled sets. Let $A \nabla B$ stand for the disjoint union of A and B and $A \times B$ for the Cartesian product.

Definition 6.3.1. *The sum of two labeled sets* A, B *is* $A \oplus B = \; < A \nabla B, l_A \oplus l_B >$ *where:* $l_A \oplus l_B(x) = l_A(x)$ *if* x *is in* A *and* $l_B(x)$ *if* x *is in* B.

(*Caveat*: actually, the definition of $l_A \oplus l_B(x)$ is slightly more complicated due to the fact that we need to take the disjoint union of A and B.)

Definition 6.3.2. *The product of two labeled sets* A, B *is* $A \otimes B = < A \times B, l_A \otimes l_B >$ *where:* $l_A \otimes l_B(x, y) = max\{l_A(x); l_B(y)\}$.

For instance, consider $E = <$ Even$, l_{Even}(x) >$ and $O = <$ Odd$, l_{Odd}(x) >$

$E \oplus O = \; < N, id_N(x) >$
$E \otimes O = \; < \{< x, y >: x$ even and y odd $, max\{id_N(x); id_N(y)\} >$

Definition 6.4. *Definition of numerosity. A numerosity function for the Class L of all countable labeled sets is a map* **num:** $L \to \mathcal{N}$ *onto a linearly ordered set $< \mathcal{N}, \leq >$ such that the following properties are satisfied:*

(1) *If* $\#A_n \leq \#B_n$ *for all n, then* **num**$(A) \leq$ **num**(B)
(2) $x <$ **num**(A) *iff* $x =$ **num**(B), *for some* $B \subset A$
(3) *If* **num**$(A) =$ **num**(A') *and* **num**$(B) =$ **num**(B') *then* **num**$(A \oplus B) =$ **num** $(A' \oplus B')$ *and similarly for* $\otimes$.

Intuitively:

(1) If all finite approximations indicate that the numerosity of A is not greater than the numerosity of B, then **num**(A) is indeed smaller than or equal to **num**(B)
(2) Proper subsets have strictly smaller numerosity
(3) Numerosities are consistent with sum and product operations on labeled sets.

Notice that $<$ Fin$, (N, \leq), \# >$ satisfies (1), (2) and (3); $<$ Sets, Card, $\| \| >$ and $<$ WO, Ord, $| | >$ do not satisfy all three properties.

Proposition 6.5. *Consequences of the axioms.* Let us assume that there is indeed such a numerosity function. It can then be shown that:

(i) $\mathcal{N}$ has a least element 0, i.e. **num(0)** (the numerosity of the empty labeled set)
(ii) All labeled singletons have the same numerosity, denoted 1 (the numerosity of the canonically labeled set $\{\mathbf{0}\}$)

(iii) Every numerosity x ($=$ **num**(A)) has a successor $x + 1$ ($=$ **num**$(A \oplus \{a\})$ where $\{a\}$ is any labeled singleton; moreover, for any labeled set A different from $\mathbf{0}$ such that **num**$(A) = x$ there is a predecessor numerosity $x - 1$

(iv) If $< A, l_A >$ is finite then **num**$(A) = \#A$, i.e. the cardinality of A.

So $\mathcal{N}$ contains a copy of the natural numbers. Moreover, on account of property (iii) one can define addition and multiplication on numerosities as follows:

$$\mathbf{num}(A) + \mathbf{num}(B) = \mathbf{num}(A \oplus B);$$
$$\mathbf{num}(A) \cdot \mathbf{num}(B) = \mathbf{num}(A \otimes B)$$

Further, one can show that $< \mathcal{N}, +, \cdot, 0, 1, \leq >$ is a positive semi-ring with neutral elements (theorem 2.3 in Benci and Di Nasso 2003; neutrality means that $a + 0 = a$ and $a \cdot 1 = a$). Finally, $\mathcal{N}$, the set of numerosities, can be shown to be embeddable in a set of hypernatural numbers and to give rise to a non standard model of analysis. In other words numerosities behave exactly like the finite numbers.

But is there a model of such a set of axioms? Yes, there is. The construction consists in taking numerosities to be equivalence classes of non-decreasing functions from the natural numbers into the natural numbers that are equivalent modulo a 'selective' (or 'Ramsey') ultrafilter on the natural number (henceforth the discussion will be restricted to ultrafilters on the natural numbers). Indeed, the existence of a numerosity function on countable sets is equivalent to the existence of a selective ultrafilter (Benci and Di Nasso 2003). It is also well known that the existence of a selective ultrafilter is independent of ZFC.

Definition of ultrafilter   *A non-empty family U of subsets of I is called an ultrafilter over I if it is closed under supersets and under finite intersections, and if for every $A \subseteq I$, either $A \in U$ or $I - A \in U$.*

If no finite subset belongs to U then U is called *nonprincipal*.

Let U be a nonprincipal ultrafilter over N. Such an ultrafilter is said to be selective if for every function $\phi : N \to N$ there exists a $D \in U$ such that ϕ restricted to D is non-decreasing.

The importance of looking at non-decreasing functions is related to the fact that the counting of sets gives rise to non-decreasing functions.

There are several equivalent definitions of such ultrafilters which justify their being called 'selective' or 'Ramsey' but that is not essential for us at the moment (see proposition 4.1 in Benci and Di Nasso 2003).

Assume U is a selective ultrafilter. Consider the U ultrapower of N,

$$\mathcal{N} = (N^N)_U = \{[\phi]_U : \phi : N \to N\}$$

where $[\phi]_U$ is the equivalence class of ϕ modulo the equivalence relation

$$\phi \approx_U \psi \text{ iff } \{n : \phi(n) = \psi(n)\} \in U.$$

We can easily see that $\leq$ can be defined in a similar way and thus $(\mathcal{N}, \leq)$ is a linearly ordered set. This allows one to prove all the properties of Definition 6.4 and thus to define plus and times accordingly.

Theorem (as in theorem 2.3): $< \mathcal{N}, +, \cdot, 0, 1, \leq >$ is a positive semi-ring with neutral elements.

We now need to tie the work on $\mathcal{N}$ to the labeled sets.
Let us focus on

$$F = \{\phi : \phi : N \rightarrow N \text{ such that } \phi \text{ is nondecreasing}\}.$$

This is not a positive semi-ring for there is no guarantee that given $x < y$ there is a unique z such that $x + z = y$. It is only a partially ordered semi-ring. However, it has the following important property:

For every labeled set A, the approximating sequence $\gamma_A : n \rightarrow \#A_n$ is in F. Moreover, every function in F is the numerosity function of some labeled set.

Definition   For every labeled set A define $\mathbf{num}(A) = [\gamma_A]_U$.

One then checks (which is by no means trivial; see theorem 4.3), using the selectivity properties of the ultrafilter, that $\mathbf{num}$ satisfies the desired axioms. For instance one proves that $\mathbf{num}$ is onto $\mathcal{N}$, by remarking that any $\phi : N \rightarrow N$ is U equivalent to some nondecreasing sequence, hence to the approximating sequence of some labeled (countable) set.

SOME COMPUTATIONS

Given that the existence of a selective ultrafilter is a highly non-constructive assumption, we do not have a very good handle on what truths about numerosities will be induced by simply postulating that we are working with an arbitrary selective ultrafilter. However, we can construct selective ultrafilters in such a way as to make sure that certain sets are in it. For instance one can show that the following conditions can be satisfied.

(a) For each $k > 0$, $\mathbf{num}(N)$ is a multiple of k
(b) For each $k > 0$, $\mathbf{num}(N)$ is a k-th power

This amounts to making true the following statements:

$$\mathbf{num}(kN) = \mathbf{num}(N)/k$$
$$\mathbf{num}(N^2) = \sqrt{\mathbf{num}(N)}$$

While Cantor's theory of cardinality collapses the size of all countable sets, the new theory discriminates between sizes of countable sets and thus seems to vindicate some of our intuitions about sizes of infinite sets. However, one can claim that the theory discriminates too much. The reason for such criticism will emerge by reflecting on two facts. First of all, everything depends on the choice of ultrafilter. Depending on whether the ultrafilter one chooses contains the even numbers or the odd numbers, it will turn out that this will affect such properties as whether the numerosity of the natural numbers is even or odd.[19] Moreover, even the equality between the size of the odd numbers and that of the even numbers will depend on whether the even numbers are defined as containing zero or not. Both problems can be illustrated by the following example. Suppose the set of even numbers is in the ultrafilter you have chosen. Now define

Even $= \{2n : n \in N\}$ and **Odd** $= \{2n - 1 : n \in N_+\}$.

In this case, letting $N = \{0, 1, 2, 3, 4, 5, \ldots\}$, the approximating functions for the two sets are:

Even: $1, 1, 2, 2, 3, 3, \ldots$
Odd : $0, 1, 1, 2, 2, 3, \ldots$

First notice that summing up the approximating sequences of even and odd we get $1, 2, 3, 4, \ldots$

which is the approximating sequence for N. Thus **num(Even)**+**num(Odd)** $=$ **num**(N).

We see that the approximating functions of the sets **Even** and **Odd** agree on the set of odd numbers (which is not in the ultrafilter) and that the approximating function for **Even** majorizes that for **Odd** on the set of even numbers. Indeed, in this case **num(Even)** $=$ **num(Odd)** $+ 1$. As a consequence **num**$(N) = [2 \cdot$ **num(Odd)**$] + 1$, that is, the numerosity of the set of natural numbers is an odd number. Of course had we chosen an ultrafilter containing the odd numbers instead of the even ones then **num**(N) would have turned out even.

We thus see that whether the even numbers have the same numerosity as the odd numbers will depend on the choice of ultrafilter. And even having fixed the choice to an ultrafilter containing all the even numbers we now observe the following. Consider **Even** $= \{2n : n \in N\}$, **Odd** $= \{2n - 1 : n \in N_+\}$, and **Even$_+$** $= \{2n : n \in N_+\}$. We get the following:

num(Even$_+$) $=$ **num(Odd)** $<$ **num(Even)**.

---

[19] I cannot refrain from recalling what Descartes said about this type of question: "We will not bother to reply to those who ask if the infinite number is even or odd or similar things since it is only those who deem that their mind is infinite who seem to have to tackle such difficulties." (Descartes, *Principes de la Philosophie*, I.26).

And this shows how sensitive these computations are to where the counting begins. One wonders whether it might be possible to modify the theory so as to make it less sensitive to such decisions.

3.7 Philosophical remarks

3.7.1 An historiographical lesson

As should be clear from the above, I believe that the recent mathematical developments should help us abandon a 'Whig' history of the concept of infinity and to make us more receptive to the complexities of the contrasting intuitions that have shaped the attempts to cope with such a recalcitrant object. I am sure every reader can adduce his favorite examples of such misreadings; I will only mention two cases. The first shows an unwarranted negative judgment toward an author's accomplishment only on account of his not having taken the Cantorian route. The second shows the tendency to assimilate previous authors to the later Cantorian accomplishments.

The first example comes from Gardies:

Maignan montre ainsi que deux infini peuvent être égaux (ce qui ce voit, dirions-nous, à ce que leurs élements peuvent être mis en bijection) ou inégaux et que, par consequent, les relations plus grand que et plus petit que gardent leur sens entre infinis. La malheur est seulement que Maignan lancé prophétiquement sur cette voie Cantorienne, choisit des exemples d'infinis dénombrables. (Gardies 1984, p. 126)

But why think that Maignan was in teleological fashion aiming toward the Cantorian solution as opposed to considering him according to his own ambitions and intuitions? It is obvious that for Maignan the comparability of infinite sets of integers was a major desideratum (and even a 'data' of intuition) that he tried, unsuccessfully, to coherently develop in his *Cursus*. But he was not unsuccessful on account of not having reached the Cantorian conclusions. Rather, his theory was unsuccessful because, as I pointed out, it was unstable. I should also remark that the kind of evaluative tendency displayed by Gardies is not without consequence when interpreting the author being studied, as witnessed by Gardies's unqualified reading of Maignan's criterion of identity. We have seen that a more charitable interpretation is needed in order to account for Maignan's discussion of Galileo's paradox.

The second case is taken from P. Duhem. In his discussion of Gregory of Rimini he says:

Grégoire de Rimini avait certainement entrevu la possibilité du système logique que M. Cantor est parvenu à construire . . . (Duhem 1955, p. 392)

Needless to say, bringing in Cantor is absolutely of no use for understanding Gregory nor does Gregory's importance lose or gain by being associated to Cantor. This list could be easily added to (see Rabinovitch 1970, Maier 1949, Bunn 1977 and other cases discussed earlier).

Naturally, I am also trying to avoid the opposite mistake. That is, I am not suggesting that we now should reread the history of infinity to show that, on account of the recent mathematical work, Ibn Qurra, Grosseteste, Maignan, Bolzano, and so forth, were 'right' all along. This would do nothing but rehearse the debate that followed on the heels of Robinson's discovery of non-standard analysis and his claim that the work of the early infinitesimalists had thus been vindicated. Such a line of argument was effectively rebutted by Bos (1974) who showed that the laws of non standard analysis do not match those displayed by Leibniz's treatment of second-order differentials. However, in both cases the new mathematical theories are of use exactly in showing that some aspects of the previous intuitions, however vague and imprecise, could be made systematic. And I believe this should not take the form of another 'Whig' history but simply of opening up the conceptual spectrum for taking seriously the multitude of intuitions that shaped the history of infinity.

While there is a sense of 'inevitability' implicit in the kind of 'Whig' history I have referred to, namely the sense that there was only one 'right' way to go when developing a theory of infinite number, no theoretical claim to such effect is explicitly given in the sources mentioned above. By contrast, in the next section we will encounter a philosophical argument due to Gödel to the effect that if one wants to extend the notion of number from the finite to the infinite there is no alternative but to accept the Cantorian notion of cardinality. We now turn to that claim.

3.7.2 Gödel's claim that Cantor's theory of size for infinite sets is inevitable

An argument that points to the inevitability of the Cantorian choice of defining number in the infinite realm has been given by Gödel (1990, pp. 254–270) in his paper "What is Cantor's Continuum Problem?" The passage is long but it is important to quote it at length:

Cantor's continuum problem is simply the question: How many points are there on a straight line in Euclidean space? An equivalent question is: How many different sets of integers do there exist?

This question, of course, could arise only after the concept of "number" had been extended to infinite sets; hence it might be doubted if this extension can be effected in a uniquely determined manner and if, therefore, the statement of the problem in the simple terms used above is justified. Closer examination, however, shows that Cantor's definition of infinite numbers really has this character of uniqueness. For whatever "number" as applied to infinite sets might mean, we certainly want it to have the property that the number of objects belonging to some class does not change if, leaving the objects the same, one changes in any way whatsoever their properties or mutual relations (e.g. their colors or their distribution in space). From this, however, it follows at once that two sets (at least two sets of changeable objects of the space-time world) will have the same cardinal number if their elements can be brought into a one to one correspondence, which is Cantor's definition of equality between numbers. (Gödel 1990, p. 254)

The two claims are connected by the following explanation:

For if there exists such a correspondence for two sets A and B it is possible (at least theoretically) to change the properties and relations of each element of A into those of the corresponding element of B, whereby A is transformed into a set completely indistinguishable from B, hence of the same cardinal number. (Gödel 1990, p. 254)

After explaining this notion of one to one correspondence for sets of physical things, and observing that it must also apply to numbers, Gödel went on to conclude:

So there is hardly any choice left but to accept Cantor's definition of equality between numbers, which can easily be extended to a definition of "greater" and "less" for infinite numbers by stipulating that the cardinal number M of a set A is to be called less than the cardinal number N of a set B if M is different from N but equal to the cardinal number of some subset of B. (Gödel 1990, p. 255)

Gödel's reflection aims at showing that in generalizing the notion of number from the finite to the infinite one inevitably ends up with the Cantorian notion of cardinal number. The key step in the argument is the premise and the theory of numerosities can help us see that the premise already contains in itself the Cantorian solution. In fact, the premise takes as evident the request that "the number of objects belonging to some class does not change if, leaving the objects the same, one changes in any way whatsoever their properties or mutual relations (e.g. their colors or their distribution in space)." While the premise constitutes no problem when dealing with finite sets, one might question its acceptability in the realm of the infinite. Indeed, in the theory of numerosities we cannot grant the premise when it comes to infinite sets. For, while it is possible to abstract from the nature of the objects themselves there is one type of relation that affects the counting, namely the way in which the elements are grouped. Such grouping makes no difference in the realm of finite sets of integers. But when we move to infinite sets a rearrangement of the grouping will in general affect the approximating functions and thus the numerosity of the set. Someone committed to the counting embodied in the theory of numerosities might thus reasonably resist accepting the premise on which Gödel bases his argument and thus also resist the claim that the generalization of number from the finite to the infinite must perforce end up with the notion of cardinal number. In short, having a different way of counting infinite sets shows that while Gödel gives voice to one plausible intuition about how to generalize "number" to infinite sets there are coherent alternatives.[20]

To the possible objection that numerosities on countable sets are not enough to provide size comparisons on larger sets (in the Cantorian sense), I would reply by pointing to the extension of the theory of numerosities to uncountable sets (see the various contributions by Benci, Di Nasso, and Forti). The above reflections dovetail quite well with some comments on Gödel's passage found in Buzaglo (2002, p. 127) and Parker (2009). However, in this case I have the advantage of pointing not only to mere possibilities but rather to actually worked-out mathematical systems of counting and numbering that apply to infinite sets and do not coincide with the Cantorian norion of "cardinal number". Buzaglo for instance asserts that "Gödel was right in claiming that certain constraints force a unique expansion of the concept of number, but it is possible to choose other constraints that are no less natural and obtain a different definition of number". Let me also point out that the the the recent works by Benci, Di Nasso, and Forti might also provide the tools for answering the question raised by Buzaglo: "Is it possible to create a forced extension of finite cardinality which distinguishes between different infinite magnitudes and yet is incommensurate with Cantor's concept of infinite cardinality? The answer might have implications for the philosophy of mathematics." (Buzaglo 2002, p. 49)

Parker (2009) also criticizes this passage by Gödel pointing out that although the principle of one to one correspondence is very intuitive also the part-whole principle is very intuitive (Parker calls the two principles "Hume's Principle" and "Euclid's Principle"):

Admittedly, Gödel gives a very compelling argument for Hume's Principle: If two sets can be put in one to one correspondence, then we could conceivably alter the individual elements of one set until they were indistinguishable from their counterparts in the other, and then surely the two sets must have the same numerosity. I say this is very compelling,

---

[20] I find it telling that in the long introduction to Gödel's (1990) paper "What is Cantor's Continuum Problem?", Gregory Moore does not critically evaluate this central claim with which Gödel opens the essay; such is the widespread acquiescence to the idea that Cantor's theory is inevitable. It should also be added that Moore summarizes Gödel's position as consisting of "the minimal requirement that if two sets have the same cardinal number then there exists a one to one correspondence between them" (Gödel 1990, p. 160). It is actually the other way around. Gödel's claim is the following: "two sets [...] will have the same cardinal number if their elements can be brought into a one to one correspondence".

but nonetheless it is only an intuition pump. Gödel disregards the fact that Euclid's Principle is *also* intuitively compelling! If set *A* contains everything that is in set *B* and also some further things, then it contains *more*. *Both* Euclid's *and* Hume's Principles seem forced on us. To have a consistent theory of transfinite numerosity, we must break free of these forces, much as Gauss and Lobachevsky broke free of the parallel postulate. We have learned from them that intuitions do not limit our freedom to form counterintuitive conceptions. Even if Hume's Principle seems stronger than Euclid's, no adequate reason has been given to believe that it is unrevisable or a brute fact. It is up to us to choose our preferred principles, or to articulate an arsenal of different concepts incorporating different principles. (Parker 2009, pp. 106–107)

Indeed, the theory of numerosities gives a concrete example of an alternative theory of counting with infinite sets of integers that is much more compelling than the weak alternative obtained by applying Duggan's theorem mentioned by Parker. But this is grist to Parker's mill.

It would be interesting here to study how the options provided by the theory of numerosities (and Katz's theory) impact, presumably favorably, both Buzaglo's work on forced (but not strongly forced) expansions and Parker's method of conceptual articulation. This would have interesting reverberations on issues of philosophy of language. But I will leave that for others to carry out.

3.7.3 Generalization, explanation, fruitfulness

So far the point has been conceptual: both CS (Katz's theory) and the theory of numerosities show the coherence of the idea of assigning different sizes to infinite sets of natural numbers in accord with the part–whole principle. The reader should not infer from what has been said that any claim is being made as to the relative mathematical fruitfulness of the theory of cardinals (or ordinals) versus the theory of numerosities. For one thing, the two are not in conflict. Conflict emerges only if both notions are taken to explicate the same intuitive notion of size. Moreover, no claim is being made here as to the mathematical fruitfulness of the theory of numerosities, notwithstanding its interest as an alternative foundation for non-standard analysis (which was the acknowledged goal of the theory according to Benci and his co-authors).

However, this once again raises the issue of when a generalization is fruitful. In this context, I would like to revisit the discussion of Bolzano and Cantor contained in Kitcher's *The Nature of Mathematical Knowledge* (1984). Kitcher was aiming at classifying patterns of mathematical change and investigating why such patterns were rational. One such pattern is generalization and, in this light, Kitcher set himself the goal of explaining why Cantor's extension of finite arithmetic to infinite sets was fruitful and rational. According to Kitcher, generalizations come cheap.[21] But the significant generalizations, the ones that truly make for outstanding mathematical achievements, are explanatory. Kitcher describes as follows the nature of explanatory generalizations:

They explain by showing us exactly how, by modifying certain rules which are constitutive of the use of some expressions of the language, we would obtain a language and a theory within which results analogous to those we have already accepted would be forthcoming. From the perspective of the new generalization, we see our old theory as a special case, one member of a family of related theories.[. . .] Those "generalizing" stipulations which fail to illuminate those areas of mathematics which have already been developed [. . .] are not rationally acceptable. (Kitcher 1984, pp. 208–209)

In order to illustrate the difference between explanatory generalizations and those that are not explanatory, Kitcher went on to compare Bolzano's and Cantor's approaches to generalizing arithmetic to infinite sets. Bolzano's attempt is judged sympathetically by Kitcher. Indeed, he even claims that Bolzano tried to stick, in the choice between Hume's Principle and Euclid's Principle, to the *more* intuitive of the two requirements:

Intuitively, it appears that the second condition [part–whole condition] is more important, so that Bolzano declares that two sets do not have the same number if one is a proper part of the other. Quite consistently, he goes on to claim that the existence of one to one correspondence between two sets is only a sufficient condition for the sets' having the same number of members when the sets are finite. Unfortunately, Bolzano's choice makes him unable to develop a theory of infinite numbers which will have analogs of standard theorems about numbers. His attempt to generalize casts no light on ordinary arithmetic, and, not surprisingly, no accepted theory of the transfinite results from his writings. Bolzano's stipulation of "sameness of size" for infinite sets fails to serve any explanatory ends, and so it is not rational to extend mathematical language by adding it. (Kitcher 1984, p. 210)

While there is no doubt that Bolzano did not develop a theory of infinite sets of numbers, Kitcher seems to imply that the failure was due to a hopeless attempt, namely preserving the part–whole principle for infinite sets. By contrast, Cantor by abandoning "the intuitive criterion for inclusion" was able to generalize the ordinary arithmetical laws. The fruitfulness of Cantor's notion of "having the same power" was displayed by his theorem that real numbers do not have the same power as the natural numbers. This was then exploited in the development of a

---

[21] I have discussed the tangle between generalization and explanation, also in connection to Kitcher, in Mancosu (2008a).

theory that provided analogues for the ordinary arithmetical operations. Thus, Kitcher says:

Unlike Bolzano's attempt, Cantor's stipulation is rationally acceptable because it provides an explanatory generalization of finite arithmetic. Note first that the ordinary notions of order among numbers, addition of numbers, multiplication of numbers, and exponentiation of numbers are extended in ways which generate theorems, analogous to those of finite arithmetic (for example, if a, b are infinite cardinals, we have the result: $(\exists x)x > a$, $a + b = b + a, a + b \geq a, a \cdot b \geq a, 2a > a$, and so forth). By contrast, because he cleaves to the intuitive idea that a set must be bigger than any of its proper subsets, Bolzano is unable to define even an order relation on infinite sets. The root of the problem is that, since he is forced to give up the thesis that the existence of one to one correspondence suffices for identity of cardinality, Bolzano has no way to compare sets with different members. (Kitcher 1984, p. 211)

But in light of the mathematical theories we have discussed, this explanation of Cantor's superiority over Bolzano should leave us puzzled. The theory of numerosities, or even Katz's theory, is able to distinguish between sizes of different sets of integers. Moreover, the theory of numerosities generalizes finite arithmetic much more thoroughly than Cantor's theory of ordinals or cardinal numbers. Indeed, all the standard algebraic laws for addition and multiplication hold for numerosities. Hence, the advantage of Cantor's theory cannot reside here. Kitcher does adduce a second consideration:

Second, Cantor's work yields a new perspective on an old subject: we have recognized the importance of one to one correspondence to cardinality; we have appreciated the difference between cardinal and ordinal number; we have recognized the special features of the ordering of natural numbers. (Kitcher 1984, p. 211)

Even here one could counter that by generalizing the process of counting to counting infinite sets, the theory of numerosities has allowed us to appreciate the importance of one aspect of counting that does not coincide with either cardinality or ordinality (once we move to the infinite). The real advantage, even for Kitcher, is not at this level. He makes the following claim on behalf of Cantor's theory:

But we do not even need to go so far into transfinite arithmetic to receive explanatory dividends. Cantor's initial results on the denumerability of the rationals and algebraic numbers, and the non-denumerability of the reals, provide us with new understanding of the differences between the real numbers and the algebraic numbers. Instead of viewing transcendental real numbers (numbers which are not the roots of polynomial equations in rational coefficients) as odd curiosities, our comprehension of them is increased when we see why algebraic numbers are the exception rather than the rule. (Kitcher 1984, p. 211)

This last kind of motivation is different from the one based on the fact that the theory generalizes the laws holding for finite arithmetic in that it brings into play the exploitation of the theory for understanding parts of mathematics that do not relate only to its ability to explain aspects of the notions (finite arithmetic) generalized by the new notion. In other words, it is the fruitfulness of Cantor's theory to mathematical practice that seems to decree its explanatory superiority with respect to a theory like the theory of numerosities. And while it might be unfair, and perhaps too early, to judge the recent theory of numerosities against Cantor's theory of cardinality, comparing the two might help us focus on the problem of mathematical fruitfulness.[22] What accounts for it? And if the account is given in terms of explanatory generalization, understanding, and similar notions how can we account for the latter? This is a problem that an interesting epistemology of mathematics should try to address.[23]

3.8 Conclusion

In this chapter my goal was to establish the simple point that comparing sizes of infinite sets of natural numbers is a legitimate conceptual possibility. I have addressed the problem of counting infinite sets from three different perspectives. I used the historical part to motivate the naturalness of the intuition that there are different sizes between infinite sets of natural numbers. The mathematical part showed that this intuition was capable of being made rigorous (without entering into claims as to whether the original intuitions were 'fully' captured). Finally, in the third part, I hope to have persuaded the reader that the possibility of comparing Cantor's theory against the alternative theories of class sizes (CS) and numerosities allows us to analyze more finely, and in some cases debunk, the arguments that claim either the inevitability of the Cantorian choice (Gödel) or

---

[22] Modulo my claim, that the work on numerosities is important on conceptual grounds and within the scope of its original aim (an alternative foundation for non-standard analysis), I fully agree with the sentiments expressed in an email sent to me by Jeremy Avigad (March 12, 2009): "For me, what would decide whether Katz's and Benci-Nasso's theories are genuinely interesting would be the depth of the ideas and how they interact with other parts of mathematics. I take mathematics to be a way of organizing and explaining our scientific experiences. At the base, it comes close to empirical activities like counting, measuring, and making predictions; but we build up theoretical edifices to make sense of these, and so the story becomes more elaborate. As you know, I think it is very important to try to understand what underlies our judgments as to whether a theory is good or not. Determining whether Cantor or Katz or Benci-Nasso have contributed something important requires a more elaborate story that goes well beyond the two starting intuitions." I should perhaps add that these reflections by Avigad were not intended as objections to the claims made in this chapter nor do I take them to be so, as I fully agree with the thoughts expressed therein.

[23] See the various contributions in Mancosu (2008b).

that account for the (alleged) explanatory nature of the Cantorian generalization by appealing to the (alleged) non-rational nature of preserving the part–whole principle. By doing so, we were able to connect the topic of this paper to questions that are now at the forefront of recent work in the philosophy of mathematics concerning issues of fruitfulness, explanation, generalization, and so forth.

4

In good company? On Hume's Principle and the assignment of numbers to infinite concepts

4.1 Introduction

In Chapter 3, we have explored the historical, mathematical, and philosophical issues related to the new theory of numerosities. The theory of numerosities provides a context in which to assign 'sizes' to infinite sets of natural numbers in such a way as to preserve the part–whole principle, namely if a set *A* is properly included in *B* then the numerosity of *A* is strictly less than the numerosity of *B*. Numerosity assignments differ from the standard assignment of size provided by Cantor's cardinality assignments. In the third part of Chapter 3, we have also probed the philosophical consequences of these new developments by analyzing an argument due to Gödel according to which the Cantorian generalization from finite to infinite numbers was 'inevitable' and Kitcher's evaluation of the Bolzano/Cantor opposition in his account of the 'rationality' of transitions between mathematical practices. That reflecting on theories of numerosities could also be useful in rethinking central issues of neo-logicism became clear to me after I read some relevant passages of Heck (2011). After having mentioned my historical and mathematical discussion concerning the assignments of numerosities to infinite sets that preserved the part–whole principle and the theory of numerosities, Heck summarized the upshot of my contribution as follows:

Mancosu's announced goal in his paper is "to establish the simple point that comparing sizes of infinite sets of natural numbers is a legitimate conceptual possibility" (Mancosu, 2009, p. 642 [now in Chapter 3 of the present book]). I think it is clear that he succeeds. But if it is conceptually possible that infinite cardinals do not obey HP [Hume's Principle], then it is conceptually possible that HP is false, which means that HP is not a conceptual truth, so HP is not implicit in ordinary arithmetical thought. (Heck 2011, p. 266)

I will come back to this argument by Heck but here I would like to flag the importance of Heck's reflection on numerosities for the developments contained in this chapter. While the appeal to the theory of numerosities is, as I will later claim, not needed for the 'good company' objection, Heck's worries on the theory of numerosities were my point of departure. It was by thinking about Heck's considerations on the matter that I was led to generalize his line of attack on the analyticity (or conceptual status) of HP to a line of thought resulting in what I call below a 'good company' objection to Hume's Principle.

In addition to the introduction and the conclusion, the chapter has six main sections. §4.2 presents some introductory discussion of neo-logicism, Hume's Principle, and Frege's theorem. §4.3 takes a historical look at nineteenth-century attributions of equality of numbers in terms of one–one correlation and argues that there was no agreement as to how to extend such determinations to infinite sets of objects. The intended effect of this section is to persuade the reader that the criterion of one–one correlation for assigning numbers to infinite sets is not to be taken for granted, either historically or systematically. This leads to §4.4 where I show that there are countably-infinite many abstraction principles that are 'good', in the sense that they share the same virtues of HP (or so I claim) and from which we can derive the axioms of second-order arithmetic. §4.5 connects this material to a debate on Finite Hume's Principle between Heck and MacBride and highlights the importance of two different projects in neo-logicism, a hermeneutic and a reconstructive project. Then, in §4.6, I go back to Heck's original objection to the analyticity of Hume's Principle based on the theory of numerosities and state the 'good company' objection as a generalization of his objection. This is followed by a discussion of the argument and by a taxonomy of possible neo-logicist responses to the 'good company' objection. Finally, §4.7 makes a foray into the relevance of this material for the issue of cross-sortal identifications for abstractions.

4.2 Neo-logicism and Hume's Principle

Let me begin by recalling a few central facts about neo-logicism. Neo-logicism is an attempt to revive Frege's program by claiming that important parts of mathematics, such as second-order arithmetic, can be shown to be analytic (or to have a status akin to analytic). The claim rests on a logico-mathematical theorem and a cluster of philosophical arguments. The theorem is called *Frege's theorem* (see Heck 2011) namely that second-order logic with a single additional axiom, known as N= or Hume's Principle (HP), deductively implies (modulo some appropriate definitions) the ordinary axioms for second-order arithmetic. The cluster of philosophical claims is related to the status (logical and epistemic) of N= or Hume's Principle. Let us recall that in a Fregean context the second-order systems have variables for concepts and objects (individuals). In addition, we have functional symbols that express functions that when applied to concepts yield objects as values. One such functional symbol[1] is # (another is the course of value operator in *Grundgesetze*, see Frege 1893). The intuitive meaning of # is as an operator that when applied to concepts outputs numbers (which are thus construed as objects) corresponding to the cardinal number of the objects falling under the concept. For instance, '$\#x:(x = $ Barack Obama)' denotes the number 1, where '$x = $ Barack Obama' expresses the concept of being identical to Barack Obama. $N =$ or Hume's Principle (HP) has the following form:

HP $(\forall B)(\forall C)\,[\#x:(Bx) = \#x:(Cx) \leftrightarrow B \approx C]$

where $B \approx C$ is short-hand for one of the many equivalent formulas of pure second-order logic expressing that "there is a one–one correlation between the objects falling under B and those falling under C".

As remarked, the right-hand side of the equivalence can be stated in the pure terminology of second-order logic. The left-hand side gives a condition of numerical (cardinal) identity for concepts. The concepts B and C have the same cardinal just in case there is a one–one correlation between the objects falling under them. In the ordinary framework of second-order logic # stands for a total function on concepts and thus all concepts have a cardinality. We have '$\#x:(x \neq x)$' which will have to denote the cardinality of a concept having empty extension (for this reason Frege uses this concept to define '0' as '$\#x:(x \neq x)$') and # might also be used to define new concepts so that it makes perfect sense to have a term such as '$\#y:(\#x:(x \neq x) = y)$' which according to Frege's definitions will denote the number 1. Notice that there will also be an object corresponding to '$\#x:(x = x)$' which will turn out to denote an infinite number (since there are at least countably-infinite many objects obtainable in the system, namely the natural numbers).[2] We will return to the issue of infinity later.

HP combines two biconditionals (see Dummett 1998). The first aims at capturing with a definition the notion that in natural language we would express with 'just as many'. The biconditional is offered as a definition in *Grundlagen*, §72:

---

[1] Frege himself does not use the symbol '#'. In what follows I will not always carefully keep track of whether # is treated as a variable-binding term operator (which applies to well-formed formulas) or whether it is treated as applied to concepts (or whatever is in the range of the monadic second-order variables). As long as we have full comprehension the two approaches are equivalent.

[2] The literature on 'anti-zero' (i.e. $\#x:(x = x)$) is quite extensive. See, among others, Hale and Wright (2001*a*), Rumfitt (2001), Clark (2004), and Heck (2011). Nothing in this chapter will be affected by what solution one embraces on anti-zero, as all the principles I will discuss are on a par on this issue.

a)   there are just as many Bs as Cs iff there is a one–one correlation between the objects falling under B and those falling under C.

This is Frege's definition of 'gleichzahlig' (equinumerous). Thus, to say that there are just as many knives as forks on the table is captured by the claim that the knives and the forks can be associated, say by matching them, so that to each fork corresponds a unique knife and to each knife a unique fork. In section 73, Frege provides us with a theorem, which gives the second biconditional:

b)   the number of Bs and the number of Cs is the same iff there are just as many Bs as Cs.

HP is usually stated as:

the number of Bs and the number of Cs is the same iff there is a one–one correlation between the objects falling under B and those falling under C.

HP is of course an easy consequence of a) and b) in Frege's development but is instead taken as an independent principle in neo-logicism.[3]

According to neo-logicism, HP is not a logical principle but it can be said to be constitutive of the concept of (cardinal) number and for this reason, among others, it is argued that it is analytic.[4] Much of the discussion on the viability of neo-logicism concerns exactly the epistemological/semantical status of HP and whether it is analytic.

Before proceeding, it is important to stress that, for the neo-logicist, the condition for identity statements between numbers (ascribed to concepts) is given by means of the right-hand side of HP, which captures the Fregean solution to determinations of cardinality (and which has much in common with the notion that in Cantor gives rise to the powers [*Mächtigkeiten*], i.e. Cantor's cardinal numbers). However, we have to question whether the right-hand side yields the required notion. For, if what the logicist and the neo-logicist are concerned with is arguing that *our ordinary knowledge* of arithmetic is based on an analytic principle and if HP, as stated, turned out not to be the principle underlying our ordinary knowledge of arithmetic, then Frege's theorem would only be of technical interest but would fail to say anything whatsoever about the nature of our knowledge of arithmetic. We will see that even this way of posing the problem is an issue but it is a good starting point.

---

[3] Lest the reader be misled, I would like to emphasize that Frege's way to HP goes through the notion of 'extension' and it is thus very different from the neo-logicist path to it.

[4] Whenever I state the neo-logicist position as a claim to the effect that HP is an analytic truth the reader should read: "HP has the status of an analytic truth or some special status akin to analytic truth".

4.3   Numerosity functions: Schröder, Peano, and Bolzano

Let us take a step back. Frege introduces, in §63 of the *Grundlagen*, the discussion of numerical identity with a reference to Hume, which is what later led Boolos to dub N= as Hume's principle. Dummett has argued that this is a misnomer and HP is a good compromise for indicating the principle.[5] In §63 Frege goes on to say: "This opinion, that numerical equality or identity must be defined in terms of one–one correlation, seems in recent years to have gained widespread acceptance among mathematicians." (Frege 1884, p. 74, tr. Austin, p. 74) The footnote refers to Schröder's *Lehrbuch der Arithmetik und Algebra* (Schröder 1873), to Kossak's *Die Elemente der Arithmetik* (Kossak 1872), and to Cantor's *Grundlagen einer allgemeinen Mannigfaltigkeitslehre* (Cantor 1883).

Kossak's statement, apparently reporting Weierstrass's definition of equality and inequality of numbers given in lectures delivered in 1865–66, is very similar to Hume's original statement concerning the equality of numbers given in terms of correspondence between the units making up the numbers (a conception of number that Frege does not accept).[6] Cantor's first work containing his definition of 'power' goes back to 1874 (see Chapter 3) but Frege, in *Grundlagen*, refers to Cantor's *Grundlagen einer allgemeinen Mannigfaltigkeitslehre* (Cantor 1883). As we know, Frege's definition of equinumerosity follows Cantor's approach in terms of one–one correlation.

Schröder's presentation is quite interesting.[7] Schröder is not defining the equality of numbers but rather when two collections have the same number. He says:

If two things have to be compared with respect to their numerosity—or, better said, two *collections* of things with respect to the number of units contained in them—the relation that must be set up between them is probably the most general that can be conceived. This seems to be indeed the case if one cannot determine anything more about its form but that it is one–one, namely that a *univocal* connection can be set between each unity of one collection and always just *one* unit of the other collection. The connection can be of any

---

[5] For the history of this debate see Heck (2011, p. xi).

[6] "Wir unterscheiden jetzt zunächst *gleiche und ungleiche Zahlen*; *gleich* sind zwei Zahlen, wenn zu jedem Element der einen Zahl ein Element der andern gehört; *ungleich* sind zwei Zahlen, wenn bei dieser Gegenüberstellung der einzelnen Elemente bei der einen Zahl ein oder mehrere Elemente übrig bleiben, und zwar heisst diese dann die *grössere* von beiden; sie enthält mehr Einheiten als die andere. Bei dieser Feststellung des Begriffes der Zahl habe ich im Wesentlichen dieselben Worte angewandt, welcher Herr *Weierstrass* sich zu diesem Zwecke in seiner in der Einleitung bezeichneten Vorlesung [Winter-Semester 1865-1866] bediente." (Kossak 1872, p. 16) Similar definitions based on one–one correlation and limited to finite numbers are found in other lectures by Weierstrass and Kronecker.

[7] I find it surprising that Schröder has hitherto not been given more attention by Frege scholars. There are passages that are strikingly similar to Frege's own approach, for instance, those in which

form whatsoever, for instance one associates one unit to the other unit by means of a mere conceptual connection.[8]

The statement so far simply says that in order to compare two collections with respect to the number of objects they contain, the notion of one–one correlation is all that is needed. Then Schröder explains by way of an example how the notion of one–one correlation can give rise to a criterion for establishing whether the elements of two collections or sets have (more literally: occur/are in) the same number [*Anzahl*]. Consider a collection of apples and one of nuts. Schröder says that the apples and nuts in the collections have the same number [*sind in gleicher Anzahl vorhanden*] if one can, through a continued association, match, one at a time, one apple to one nut—say by laying them next to each other—and end up without any unmatched apples or nuts.

The general definition is then given as follows:

(Definition 1) If it is possible to associate with each single thing of a kind one of a different kind so that there are no things left over of either kind, then we say: the things of the first kind have the same number [*sind in derselben Anzahl*] as those of the other kind.[9]

Schröder points out that a determination of number is always relative to a concept and the one where he speaks about the generality of arithmetic in comparison to geometry. Frege himself acknowledged Schröder's ideas on the first point but added: "What is true in this [Schröder's] account is wrapped up in such distorted and misleading language, that we are obliged to straighten it out and sort the wheat from the chaff." (*Grundlagen*, §51). As for the comparison between the applicability of geometry and arithmetic Schröder says: "Bei dem Abbildungsverfahren des Zählens ist daher die Abstraktion die grösste, und muss dieses somit einerseits die grösste Vereinfachung der Vorstellungen nach sich ziehen, andrerseits aber wird deshalb auch die Anwendbarkeit dieses Abbildungsverfahrens eine weitergehende sein: das Gebiet des zählbaren ist viel umfassender als z.B. dasjenige des plastisch darstellbaren" (Schröder 1873, p. 6). Of course, by emphasizing the interesting similarities, I do not intend to downplay the very different philosophy of number (and concepts) that distinguishes Frege and Schröder. Two forthcoming publications, Textor (forthcoming) and Tappenden (forthcoming), discuss some interesting aspects of the relation between Schröder's *Lehrbuch* and Frege's *Grundlagen* and other works. In particular, Tappenden emphasizes the Riemannian origins of some of the terminology common to Schröder and Frege.

   8 "Wenn nur zweierlei Dinge hinsichtlich ihrer Häufigkeit—oder, besser gesagt, zwei *Mengen* von Dingen hinsichtlich der Anzahl der in ihnen enthaltenen Einheiten—verglichen werden sollen, so ist die zwischen ihnen herzustellende Verbindung wohl die allgemeinste, die sich denken lässt. Es scheint dies in der That dann der Fall zu sein, wenn über die Art und Weise derselben nichts weiter festgesetzt wird, als dass sie eine *eindeutige* sei, dass nämlich unzweideutig zwischen einer jeden Einheit der einen Menge und immer nur *einer* Einheit der andern Menge, überhaupt eine Verbindung—welcher Art sie auch sein möge—festgelegt, z. B. die eine Einheit durch eine blosse Gedankenverbindung der andern zugeordnet werde." (Schröder 1873, pp. 7–8).

   9 "Wenn es möglich ist, von Dingen einer Art ein jedes einzeln so zu verknüpfen mit einem Dinge von andrer Art, dass weder von der einen noch von der andern Art Dinge übrig bleiben, so sagen wir: die Dinge der einen Art sind in derselben Anzahl vorhanden, als die der andern Art." (Schröder 1873, p. 8).

Thus: the members of two collections of objects *A* and *B* have the same *Anzahl* if there is a one–one correlation between them.[10] The definition of inequality between the *Anzahl* of the members of two different collections is now given as follows:

If, however, the aforementioned process of juxtaposition or that association comes to an end [*nicht weiter fortgesetzt werden kann*] with the things of one kind having been exhausted but with still unassociated things of the other kind left over, then the latter things are said to be *more numerous* than the former and thus their respective numbers must be different.[11]

Did Schröder mean to include determinations of *Anzahl* between infinite sets? From the following passage we see that for infinite sets the criterion cannot be used. The clarification is essential for the goals of this chapter:

One can also conceive of the case in which the pairing relation or ideal association of both [kinds of] units can be continued without end. Then the number of the two [kinds of] objects is said to be *unbounded* and the question concerning the equality of the two numbers [*Anzahlen*] must for the time being [*vorerst*] be left undecided.[12]

Schröder continues by offering what he claims is a reformulation of the formal definition:

(Definition 2) Of two sets [*Mengen*] it will be said that they contain the same number of elements [*Einheiten*][13] when the elements of one set can be paired with those of the other set so that nothing is left over.[14]

It is clear that Schröder's attempt at characterizing the equality of the numbers of elements of two collections is aimed at finite collections and extends to infinite collections only for the comparison between a finite and an infinite collection.

---

[10] The reader should pay attention to the fact that here Schröder does not assign a number to the collections treated as a single object but rather to the collection treated as a plurality.

[11] "Wenn aber, sobald das obige Verfahren des Nebeneinanderlegens oder jener Verknüpfungsprocess nicht weiter fortgesetzt werden kann, indem sich von der einen Art keine Dinge mehr vorfinden, noch Dinge der andern Art unverknüpft geblieben sind, so heissen die letzteren *zahlreicher* als die ersteren und ihre Anzahl wird dann beiderseits eine verschiedene sein müssen." (Schröder 1873, p. 8).

[12] "Denkbar ist auch noch der Fall, dass jener Process der paarweisen Verknüpfung oder ideellen Association von beiderlei Einheiten ohne Ende fortgesetzt werden kann. Alsdann heisst die Anzahl von beiderlei Gegenständen eine *unbegrenzte* und muss die Frage nach der Gleichheit der beiden Anzahlen vorerst unentschieden gelassen werden." (Schröder, 1873, p. 8) In translating 'Menge' as 'set', I do not intend to use 'set' in the technical sense used nowadays in set theory.

[13] 'Units' would be a more appropriate translation but I will use 'elements', also in the following translations from Schröder, to increase readability.

[14] "Von zwei Mengen soll gesagt werden, dass sie gleich viele Einheiten enthalten, wenn die Einheiten der einen Menge sich denen der andern Menge, ohne dass ein Rest bleibt, einzeln zugesellen lassen." (Schröder 1873, p. 8).

This is obvious from his explicit declaration about leaving undecided the cases of comparison between infinite collections. Further confirmation is provided by Schröder's considerations in later sections of the book. In section 12, Schröder introduces the problem of proving that the ascription of *Anzahl* is independent of the ordering with which the collection of objects is given. It is exactly this aspect of Schröder's treatment that earned him the praise of his contemporaries, such as Helmholtz who, in *Counting and Measuring from an Epistemological Point of View* (1887), wrote:

Among more recent arithmeticians, E. Schröder has also essentially attached himself to the Grassmann brothers, but in a few important discussions he has gone still deeper. As long as earlier arithmeticians habitually took the ultimate concept of number to be that of a cardinal number [*Anzahl*] of objects, they could not wholly free themselves from the laws of the behaviour of these objects, and they simply took it to be a fact that the cardinal number of a group of objects is ascertainable independent of the order in which they are numbered. To my knowledge, Mr. Schröder [Schröder, 1873, p. 14] was the first to recognize that here a problem lies concealed: he also acknowledged—in my opinion justly—that there lies a task here for psychology, while on the other hand those empirical properties should be defined which the objects must have in order to be enumerable. (Ewald 1996, vol. 2, p. 729)[15]

However, Schröder stated the result as an invariance for counting with finite collections and then, in §14, he gave an argument to show that number equality and inequality, defined in §7, necessarily exclude each other. Schröder proved this by using a proof that the mathematician Lüroth had communicated to him. The theorem obviously holds (as Schröder is aware) only for finite collections:

Theorem. If the elements of a set *A* of objects can be associated in a one–one way with the elements of a set *B* so that no object is left over (if thus the elements are present in "the same number [*Anzahl*]") then there is no possible one–one association between them in which some entities are left over, i.e. every other one–one association must not leave any entities left over.

If between the elements of a set *A* and those of a set *B* there is a one–one association in which some elements are left over then every other one–one association must leave

---

[15] "An die Brüder *Grassmann* hat sich unter den neueren Arithmetikern auch Hr. E. *Schroeder* im Wesentlichen angeschlossen, ist aber in einigen wichtigen Erörterungen noch tiefer gegangen. Während die früheren Arithmetiker den letzten Begriff der Zahl als den einer Anzahl von Gegen-ständen aufzufassen pflegten, konnten sie sich nicht ganz von den Gesetzen des Verhaltens dieser Gegenstände loslösen, und sie nahmen es einfach als Thatsache, dass die Anzahl einer Gruppe von Objekten unabhängig von der Reihenfolge, in der man sie zählt, zu finden ist. Hr. Schroeder ist, so viel ich gefunden habe, der Erste, welcher erkannt hat (l. c. S. 14), dass hierin ein Problem verborgen ist: auch hat er, meines Erachtens mit Recht, anerkannt, dass hier eine Aufgabe der Psychologie vorliegt, und andererseits die empirischen Eigenschaften zu definiren wären, welche den Objecten zukommen müssen, damit sie zählbar seien." (Helmholtz 1887).

some elements left over, i.e. no such association between those elements is possible without elements left over and they are thus present in "unequal number [*Anzahl*]"[16]

At one point in the proof, Schröder remarks that he is presupposing that the set A is finite.[17]

Let us summarize Schröder's position. Schröder provides us with an analysis of the informal notion of 'having the same number' or 'there are just as many', for sets or collections. The formal definition is given in terms of the notion of one–one correlation between the elements in the sets or collections. However, the explicit aim is finite collections, for the condition is eventually spelled out, in our terminology, in terms not only of one–one correlation between the elements of the collections but rather, using a more contemporary terminology, in the following stronger terms:

The elements of A and those of B have the same number if and only if there is a one to one and onto mapping between the elements of A and those of B and every one to one mapping between them is onto.

This obviously leaves comparison between infinite sets out of consideration. Indeed, the first part of the theorem stated above fails for two countably-infinite sets: one can give a one to one mapping, say $2n \rightarrow n$, between the even numbers and the natural numbers so that no numbers are left over. But there is also a one to one mapping, say $2n \rightarrow 2n$, with infinitely many numbers left over (the odd numbers). The same example shows that the second part of the theorem also fails for infinite sets.

The theorem holds trivially for comparisons between the elements of a finite set A and those of an infinite set B. Every one to one mapping from A into B will always leave some objects of B out of the range of the mapping.

Should we conclude, in this latter case, that the number of Bs is greater than the number of As? Let me for simplicity of comparison with later developments

---

[16] "Grundsatz. Wenn sich die Einheiten einer Menge A von Objecten so eindeutig verknüpfen lassen mit den Einheiten einer andern Menge B, dass kein Rest übrig bleibt (wenn also beiderlei Einheiten "in gleicher Anzahl" vorhanden sind), so ist keine eindeutige Verknüpfung zwischen ihnen möglich, bei der ein Rest bliebe, es muss also auch jede andere eindeutige Verknüpfung ohne Rest aufgehen.

Wenn zwischen den Einheiten einer Menge A und denen einer andern Menge B eine solche eindeutige Verknüpfung existirt, bei der ein Rest übrig ist, so muss auch jede andere eindeutige Verknüpfung zwischen denselben einen Rest lassen—eine solche Verknüpfung ohne Rest ist nicht zwischen jenen Einheiten möglich, und sind dieselben "in ungleicher Anzahl" vorhanden." (Schröder 1873, p. 19).

[17] In proving invariance of counting, other authors presuppose the finiteness of the collections without making this explicit. See for instance Stolz (1885) and Bettazzi (1887). The latter is criticized in Peano (1891) in this connection.

extend Schröder's terminology to assigning numbers not only to the elements of a collection but also to the collection itself considered as an object, something that Schröder himself normally avoids. Then we can now ask: According to Schröder, is the number of the natural numbers greater than the number of {0, 1}? This determination really depends on what Schröder meant by saying that infinite sets have 'unbounded' number. There are several options compatible with what Schröder says.

1) Infinite collections do not have a number. In this case only finite sets would have a number. This is a position espoused by Dedekind in his *Was sind und was sollen die Zahlen* (Dedekind 1888), where in the note to section 161, he explicitly restricts (*der Deutlichkeit und Einfachkeit wegen*) determinations of cardinality to finite sets. In this case we cannot properly say that the number of natural numbers is greater than that of {0, 1}. We can only say that the natural numbers are 'infinite' or 'unbounded'. As a consequence, different infinite collections would also not be comparable as to 'size' (i.e. equality and inequality cannot be applied to them). This position was defended by Galileo in *Two New Sciences* and foreshadowed by Oresme and Albert of Saxony (see Chapter 3).

2) Or perhaps Schröder intended to accept a single number [∞] for all infinite sets, so that all infinite sets would be assigned the same number but no other comparison of size between them could be determined? In light of his explicit comment that infinite collections are 'unbounded' and decreeing that comparison of sizes could not be carried out on infinite sets, it is unlikely that Schröder was contemplating assigning a single number to infinite sets. Thus, I exclude that assigning a single infinite number to all infinite sets was Schröder's position but I will argue below that Peano held such a position (at least for a while).

3) A third possibility is that infinite collections have a number but we cannot determine anything at all about how they relate amongst each other or in comparison with finite sets. Again, this does not seem to be Schröder's position but it is a logical option.

In short, independently of Cantor and before Frege proposed his criterion of 'equinumerosity' for two concepts in the *Grundlagen*, we already have an attempt at spelling out, in the context of the foundations of arithmetic, ascription of cardinal numbers to the elements of collections using the criterion of one–one correlation. Since the proposal is limited to finite collections and extends at most to comparisons between finite and infinite collections, this criterion is not Cantor's. While not excluding from the outset considerations of infinite sets, Schröder's proposal seems firmly grounded on exclusive considerations of finite sets and the most natural interpretation of his approach seems to ascribe no cardinality to infinite sets. Finally, let me point out that Schröder's language remains relational (two collections 'sind in derselben Anzahl') without explicitly committing either to treating collections as single objects (and thus, to use Russell's terminology, cardinality is assigned to a collection as many but not to a collection as one) or to construing cardinalities as objects (even though at times he uses constructions such as 'the number of the collection A'). This is also why I think that of the three interpretations outlined above, only the first one is faithful to Schröder.

An interesting take on non-Cantorian assignments of cardinalities is encountered in Giuseppe Peano's work. Peano wrote an article titled "Sul concetto di numero [On the concept of number]" (1891) in which he took exception to an article by Rodolfo Bettazzi (also titled "On the concept of number" and published in 1887) in which Bettazzi had claimed that discrete quantities are such that if there is a one to one and onto correspondence between them then every one to one correspondence between them is also onto. Peano objected that this holds only for finite sets and detected a *petitio principii* in Bettazzi's approach because the concept of number is already presupposed, according to him, in that of correspondence (for, he explains, the concept of correspondence must be based on the concept of finite set). Here is Peano's objection:

The same shortcomings are present in the article of the same author "Sul concetto di numero" (Periodico di matematica per l'insegnam. secondario, anno II). Indeed, in section 19, the author wants to obtain the positive whole numbers by considering *discrete quantities*, namely those that *result from aggregates of objects considered as equal*, and he asserts that if to every object of an aggregate A one can associate [*collegare*] one and only one object of the class B, and vice-versa, it cannot happen that one can have a correspondence that associates to every object of A one and only one object of B so that there are objects left over in B. Now this proposition only holds for classes containing a *finite number* of elements, for one can associate, for instance, the integers (class B) with their doubles (class A), and the class A is contained in B, without being equal to it. And examining the proof given by the author, one sees that he appeals to theorems on permutations that hold only in the case of finitely many objects.[18]

___

[18] "Presenta gli stessi inconvenienti lo scritto dello stesso A., Sul concetto di numero (Periodico di matematica per l'insegnam. Secondario, anno II). Invero, al n. 19, volendosi ottenere il numero intero positivo colla considerazione di *grandezze discrete,* cioè quelle che *risultano da aggregati di oggetti considerati uguali,* afferma che se ad ogni oggetto dell'aggregato A si può collegarne uno ed uno solo della classe B, e viceversa, non può avvenire che si possa far corrispondere [ad] ogni oggetto di A uno solo e distinto in B, in modo che ne avanzino in B.

Ora siffatta proposizione vale solo per le classi contenenti un *numero finito* di elementi, poichè per es. si può far corrispondere ai numeri interi (classe B) i loro doppi (classe A), e la classe A è contenuta in B, senza esserle uguale. Ed esaminando la dimostrazione dell'A. si vede che si ricorre

In the same article, Peano connected the definition of 'having the same number of objects' for classes to that of one to one correspondence by giving first the explicit definition of a function $num(a)$ which yields values in the natural numbers (including zero) or a single value 'infinity' denoted by the symbol ∞.

The definition $num(a)$, for a class, is given as follows:[19]

$num(a) = 0$ iff a is the empty class;

$num(a) = n$ iff a is not the empty class and for any element x in a, $num(a\text{-}\{x\})$ $= n - 1$;

$num(a) = \infty$ iff for every n in the natural numbers (including zero), $n \neq num(a)$.

After proving several results for finite classes, paying great attention to whether the result also held for 0 and ∞, he then stated the following unrestricted principle:

With a and b two classes, if there exists a relation f that associates every element of a with an element of b and if there exists a relation g that associates with every element of b an element of a, then the classes a and b contain the same number of objects. If the classes are finite the converse holds as well.[20]

The relations in question are implicitly assumed to be one–one. Then Peano goes on to mention that if one defines two classes to have the same number iff there exists f and g as above, then one obtains the notion of Cantorian cardinality so that, in contrast to his $num(x)$ function, two infinite sets might have different numbers associated to them. However, throughout the 1890s Peano will favor his num function as the right way to capture the numerosity of sets and only in 1899 will he accept Cantorian cardinality as the right generalization of the number concept.[21] We see here Peano holding a somewhat intermediate position between Schröder and Cantor. While the first $num(x)$ function allows for one single value

a proposizioni su permutazioni, che sussistono solo trattandosi di oggetti in numero finito." (Peano 1891, p. 257, note 1).

[19] Burali-Forti (1894*b*, pp. 104–105), also followed Peano in defining a function *num* on classes in which *num* has values in the natural numbers and one additional value 'infinity'.

[20] "Essendo *a*, *b* due classi, se esiste una relazione *f* che ad ogni *a* fa corrispondere un *b*, e se esiste una relazione *g* che ad ogni *b* fa corrispondere un *a*, allora le classi *a* e *b* contengono lo stesso numero di oggetti. Sussiste pure la proposizione inversa, supposto che le classi considerate siano finite." (Peano 1891, p. 260).

[21] Peano's trajectory on this matter is interesting. By looking chronologically at Peano's writings one notes the following development:

1891 [Sul concetto di numero]: Peano defines a function on classes, *num*, that takes values in 0, the natural numbers and ∞. In the second part of his essay on the concept of number, he mentions the possibility of a different conception of infinite number given by Cantor.

1895 [Formulaire]: *num* is still as defined in 1891 [section V of the Formulaire] but in section VI one also finds the development of the theory of Cantorian cardinalities.

when applied to infinite classes, in his later position we have the espousal of the equivalence between equality of numbers and one–one correlation.

I conclude this first part of the overview by mentioning that Otto Stolz in 1885 also uses one–one correlation to compare 'multitudes' (*Vielheiten*).[22] However, his treatment is implicitly restricted to finite sets and infinite sets are not even mentioned. In fact, the proof of invariance given by Stolz is correct only if one restricts the proof to finite sets. Incidentally, also Husserl in *Philosophie der Arithmetik* (1891) when discussing such matters restricts his attention to finite sets (and so do, as we have seen, Kronecker and Weierstrass).

My third and last author of interest is Bernard Bolzano. I left him last despite his being chronologically prior to the others we have looked at, because discussing his work will facilitate the connection to the theory of numerosities, i.e. the theory that preserves the part–whole axiom. Since this topic has been discussed extensively in Chapter 3, I will be brief. Let us spell out to begin with what consequences Hume's Principle has on ascription of sizes to infinite sets of natural numbers. First of all, given that from HP + second-order logic, one can prove the existence of countably-infinite many objects, say those falling under the concept of natural number $N(x)$, one has a number $\#x{:}(Nx)$. Moreover, the theory can trivially show that every two countably-infinite sets of natural numbers have the same number.

We have already seen that Schröder's account would have excluded such a solution, for determinations of equality and inequality between the 'Häufigkeit" [numerosity] of infinite sets were explicitly excluded by him and had to be left "unentschieden". Dedekind might also have balked at this ascription of cardinality to infinite sets. As for Peano, his *num(x)* function identifies in size all infinite sets (and a fortiori all countably-infinite sets). HP is of course modeled on Cantor's equivalence for cardinal numbers and thus, when we restrict attention to countable sets, the solutions proposed by Cantor, Frege, and Peano coincide.

1898 [Formulaire]: Here the separation is more evident than in 1895 in that one has the function *num* in section V and Cantorian cardinality is denoted as *Nc'*. In a footnote, Peano says that *num* and *Nc'* are similar notions but are not identical.

1899 [Formulaire]: Peano abandons the distinction between *num* and *Nc'* and replaces them by a single notion *Num* (upper case 'N' to distinguish this from '*num*'). *Num* coincides with Cantorian cardinality.

From 1899 onwards, Peano accepts Cantor's notion of cardinal number as the right generalization of the number concept. His early resistance and his insisting on the distinction between *num* and *Nc'* during the last decade of the nineteenth century is however indicative of the conceptual struggle of coming to terms with different intuitions about the assignment of numerosities to infinite sets.

[22] "Zwei Vielheiten heissen einander gleich, wenn sich jedem Dinge der ersteren je eines der letzeren zuorden lässt und keines von dieser unverbunden bleibt. Grösser von zwei Vielheiten heisst diejenige, von welcher, nachdem jedes Ding der anderen (kleineren) je einem von ihr zugeordnet ist, noch einige Dinge (ein Rest) unverbunden bleiben." (Stolz 1885, p. 9).

Peano's solution obviously clashes with Cantor's when the comparison is between a countable and an uncountable set. But the most powerful dissenting voice in this direction was that of Bolzano, for in his case there was no indecision about the existence of different sizes (*'Vielheiten'*) for infinite sets but at the same time Bolzano rejected the criterion of one–one correlation for identity of *'Vielheiten'*.

In *Paradoxes of the Infinite* (PU, Bolzano, 1851; 1975*a*), Bolzano explicitly entertains and rejects what Frege later captures under the definition given in *Grundlagen* §72 that I cited earlier. In sections 19 to 24, Bolzano makes several important claims. In section 19 he explicitly states that infinite multitudes can differ as to their *'Vielheit'*, a term that can be translated as 'plurality', 'multiplicity' or 'numerosity'. I will stick to the first one.

Even with the examples of the infinite considered so far it could not escape our notice that not all infinite multitudes [*Mengen*] are to be regarded as *equal to one another in respect of their plurality* [*Vielheit*], but that some of them are *greater* (or *smaller*) than others, i.e. another multitude is contained as a part in one multitude (or on the contrary one multitude occurs in another as a mere part). This also is a claim which sounds to many *paradoxical*. (English translation by Russ from Russ 2005, p. 614)

In section 20, Bolzano observes that it can happen that two infinite multitudes [*Mengen*] (notice that he does not say all) can be put in what we would call a one-one correlation. He provides two examples, the first of which is a one to one and onto mapping between the interval [0, 5] and the interval [5, 12].

In section 21, he goes on to affirm that from the existence of such a mapping between infinite multitudes one can "in no way conclude *that these two multitudes [Menge] are equal to one another if they are infinite* with respect to the plurality [*Vielheit*] of their parts (i.e. if we disregard all differences between them)". In the same section he claims that relationships of equality and inequality can be meaningfully ascribed to infinite multitudes even though he leaves the criterion fuzzy:

But rather they are able, in spite of that relationship between them that is the same for both of them, to have a relationship of inequality in their pluralities [*Vielheiten*], so that one of them can be presented as a whole, of which the other is a part. An equality of these pluralities [[*Vielheiten*; Russ here translates *multiplicities* changing the meaning of the text]] may only be concluded if some other reason is added, such as that both multitudes have exactly the same determining grounds [*Bestimmungsgründe*], e.g. they have exactly the same way of being formed [*Entstehungsweise*]. (English translation by Russ from Russ, 2005, p. 617)

It has proved quite difficult to spell out what Bolzano has in mind here, for it seems that at this stage of his thinking, namely in the PU and in the *Grössenlehre* (GL, Bolzano, 1975*b*), he was ready, in contrast to what he says in the *Wissenschaftslehre*

(WL, Bolzano, 1837), to accept that the natural numbers have the same *Vielheit* as the squares of natural numbers. Bolzano does not speak here of *Anzahl* but of *Vielheit*, though the two concepts coincide for finite collections as is clear from the next section.

In section 22, Bolzano diagnosed the paradox as emerging from an unjustified extrapolation from the finite case to the infinite case. In the finite case the presence of a one–one correlation between two multitudes is sufficient to decree their equality with respect to their plurality. Explaining why this is the case for finite multitudes, Bolzano designates with 1, 2 etc. the objects of a multitude A and speaks of the last number (there must be one such, for the multitude is finite) as the number [*Anzahl*] of the multitude. Then if there is a one–one correlation between A and B one can now transfer the numbering from A to B and thus B will have the same number of elements as A. He concludes:

Therefore both multitudes have one and the same plurality, as one can also say, *equal* plurality. Obviously this conclusion becomes void as soon as the multitude of things in A is an *infinite* multitude, for now not only do we never reach, by counting, the last thing in A, but rather, by virtue of the definition of an infinite multitude, in itself there is no *last* thing in A, i.e. however many have already been designated, there are always others to designate. (English translation by Russ from Russ 2005, p. 618)

It is thus obvious that the *reason* (see §23) that allows us to infer the equality as to plurality of two finite sets is that counting them individually yields the same number. Bolzano's project cannot be confused with one that attempts to introduce the numbers in a Fregean way using the notion of one–one correlation. He simply takes the two properties (being finite and having a last element that can be numbered with a finite number) as co-extensional where the notion of finite number is primary and must be appealed to in order to prove the equality with respect to plurality of the two finite sets. In this sense his approach resembles more that of Husserl and Couturat (Husserl in *Die Philosophie der Arithmetik* restricts himself to finite numbers whereas Couturat in his book *De l'infini mathématique* (1896) postulates the existence of infinite numbers which are appealed to explain the existence of the one to one correspondence in the cases where it subsists).

While the criterion for equality of 'pluralities' is left vague in the infinite case, there is no doubt that at the time of WL Bolzano thought that many infinite countable sets have different 'pluralities' (he called them 'Weite' in §102 of the *Wissenschaftlehre*; see Chapter 3 for the passages).

Bolzano's divergence from Frege and Cantor is remarkable. For although some of the details remain fuzzy, there is no question that he accepts the criterion of one–one correlation for equality of '*Vielheiten*' only in the finite case and that he is committed to satisfying some form or other of the part–whole principle. While in WL he was obviously ready to accept differences in *Vielheiten* between what we would now call infinite-countable sets, in his later works the situation is crystal clear only for uncountable sets such as intervals of points on the real line but it is less clear for countable sets.[23] Regardless, Bolzano did not accept one–one correlation as a sufficient criterion for equality of 'plurality' and wanted to preserve the part–whole principle at least for some classes of infinite sets.

However, as we have seen in Chapter 3, nowadays we have a theory of numerosities that can preserve the part–whole principle according to which if A is a proper subclass of B then the numerosity of A is strictly less than the numerosity of B. This allows for the determination of different numerosities between the set of natural numbers, the even numbers, the prime numbers, etc.; indeed the numerosities constitute a total ordering measuring all the subsets of natural numbers, finite or infinite. The biconditional required for stating the equality of numerosities could not be given in purely logical terms (unlike the case for HP and several other principles to be presented below), for the possibility of comparing collections of natural numbers (or larger sets) according to the theory of numerosities requires, in order to define the right-hand side of the biconditional, an appeal to Ramsey ultrafilters. The existence of Ramsey ultrafilters is independent of ZFC and thus in order to define the equivalence relation required for constructing numerosities one must appeal, at least *prima facie*, to mathematical notions whose existence go beyond pure logic.[24] But the expressibility in logic of the equivalence relation required for defining identities of numerosities will not be relevant for many of the considerations to follow.

The only important thing about this theory that needs to be kept in mind for what follows is that it acts as a refinement of Cantor's notion of cardinality. In other words, one can see Cantor's work as having established the properties of an interesting equivalence relation $\approx_C$ between sets and the theory of numerosities as having established the existence of a different equivalence relation, $\approx_{PW}$, between sets (PW for Part–Whole). Limiting our focus now to sets of natural numbers the two relations satisfy the following properties:

If A and B are both finite then $A \approx_C B$ iff $A \approx_{PW} B$.
If A is finite and B is infinite then
$\neg(A \approx_C B)$ and
$\neg(A \approx_{PW} B)$
If A and B are both infinite: if $A \approx_{PW} B$ then $A \approx_C B$ but the converse does not hold.

---

[23] Bolzano obviously does not have the countable/uncountable distinction.
[24] Let me remind the reader that the existence of a numerosity function for countable labeled sets is equivalent to the existence of a Ramsey ultrafilter (see Benci and Di Nasso 2003).

Considered merely as formal equivalence relations between collections of natural numbers the above properties are compatible. If however, they are meant to spell out the notion of 'there are just as many As as Bs' that underlies our intuitive grasp of cardinality, then they cannot both be right, for they are in conflict with each other on infinite sets.

It was exactly this possibility of assigning cardinalities in a non-Cantorian way that led Heck (2011) to propose the argument against the analyticity of HP (or its status as conceptual truth) that I mentioned at the outset. We will come back later to Heck's argument and the next section will be instrumental in developing a framework which will allow us to place Heck's worries in a more general setting. Hopefully, the developments contained in this section have had the effect of convincing the reader that the extension of the criterion of one–one correlation to infinite sets cannot be taken for granted, either historically or systematically.

I will now propose a plethora of abstraction principles that are stated in purely logical terms and share HP's virtues.

4.4 A plethora of good abstractions

The above developments call for a qualification to Dummett's claim that "By the time that Frege wrote *Grundlagen* (1884), the definition [of equinumerosity in terms of one–one correspondence] had already become a piece of mathematical orthodoxy" (Dummett 1991, p. 142). Unless Dummett's remark is qualified by restricting it to finite sets, we have seen that there was no orthodoxy as to whether the criterion was extendible to infinite sets. Bolzano, Schröder, Stolz, Bettazzi, Husserl, Weierstrass, Kronecker, and Peano, show that the orthodoxy was not there before or even after Frege wrote *Grundlagen*.

But my interest in these developments is not merely historical, for I think that reflecting on the historical development can provide interesting considerations concerning the nature of the neo-logicist program and the status of HP.

I could proceed by providing formal reconstructions of Schröder, Peano, Bolzano and Frege but that would slow us down unnecessarily. Part of the problem with reconstructing the specific proposals is that in some cases, such as Bolzano's, they are not precise enough, and in other cases, such as Schröder's, they would require a formalization that would force us to abandon the standard neo-logicist framework. For instance, in Schröder's case, the standard form of using an abstraction operator, say #, could not do justice to Schröder, because the # operator automatically assigns an object to any concept (equivalently, to any extension of that concept).[25] But in Schröder, we have no infinite number corresponding to an infinite collection.[26] We have two alternatives. We can either express the principle using a logic which allows for non-denoting terms (thereby using a free logic) or we can replace the use of a # operator with a relation $R_\#(C, x)$ holding between concepts and objects.[27]

Rather than pursue the technicalities of such alternative frameworks, what I will do, and this is in any case instrumental to the later discussion, is to propose a countable infinity of abstraction principles, at least some of which can be seen as capturing some of the variety of historical developments discussed so far. Unsurprisingly, the first one is

HP: $(\forall B)(\forall C) \; [\#x{:}(Bx) = \#x{:}(Cx) \leftrightarrow B \approx C]$

The second principle captures Peano's *num* function.[28] Let Fin(C) be any of the equivalent, purely logical, second-order predicates expressing the Dedekind-finiteness of the concept C. Let Inf(C) be $\neg$ Fin(C).

PP: $(\forall B)(\forall C) \; [\clubsuit x{:}(Bx) = \clubsuit x{:}(Cx) \leftrightarrow [(\text{Fin}(B) \; \& \; \text{Fin}(C) \; \& \; B \approx C) \vee (\text{Inf}(B) \; \& \; \text{Inf}(C))]]$

The abstraction is purely logical and it is justified since (Fin(B) & Fin(C) & $B \approx C$) $\vee$ (Inf(B) & Inf(C)) is an equivalence relation on the totality of concepts.

There are easy ways to refine the Peano abstraction principle and gain an abstraction principle that is in essence due to Boolos. Boolos used model-theoretic constructions to show some proof-theoretic results concerning the strength of

---

[25] I am of course aware that the neo-logicist literature contains many proposals for blocking the totality of the # operator. My point is simply that the default choice (more in line with Frege's original approach) is to have a total operator and any deviation must be carefully argued for.

[26] For this reason, even principles such as Weak Hume's Principle (WHP), stated in MacBride (2000) and studied in detail by Heck do not quite capture Schröder's position. The principle says: (WHP) If A and B are both finite, then $\#x{:}(Ax) = \#x{:}(Bx) \leftrightarrow A \approx B$. However, since the # operator is total, the infinite concepts are assigned a cardinality, even though the principle does not license asserting anything specific about the cardinalities of infinite concepts. A further complication with Schröder's original text, as I have already pointed out, is that he does not treat collections as single objects and thus seems to assign cardinal numbers to collections as many and not to collections as one.

[27] Both approaches have been amply discussed in the literature. For a discussion of the free-logic approach see Shapiro and Weir (2000) and Tennant (2013). The "relational" approach has the advantage of distinguishing cleanly between the identity conditions and the question of existence. Boolos used the approach in "The Consistency of Frege's *Foundations*" (Boolos 1987) and also in "Is Hume's Principle Analytic?" (Boolos 1997).

[28] Cook (2012, p. 675, note 3) calls this principle FHP, which is somewhat misleading given the previous use of HPF for 'Finite Hume's Principle' in Heck (1997a). In any case, Cook's FHP is what I am calling PP. Cook also does not worry about the fact that, although not difficult, one needs to eliminate the uses of HP when working only with PP in deriving the axioms of PA_2.

different theories of second-order arithmetic and correlated Fregean theories (to be discussed below). Let us assign the cardinal numbers as usual to finite concepts by means of one–one correlation. However, for infinite concepts A and B we will assign the same number a (different from every natural number) iff A and B are both co-finite and a number b (different from a and from every natural number) iff the negation of both concepts is infinite. The notion of co-finite can be stated in pure second-order logic since $\text{Cof}(B)$ is by definition $\text{Fin}(\neg B)$. Let's formalize this principle and name it BP.

BP: $(\forall B)(\forall C)\ [!x{:}(Bx) = !x{:}(Cx) \leftrightarrow ([\text{Fin}(B)\ \&\ \text{Fin}(C)\ \&\ B \approx C] \vee [\text{Inf}(B)\ \&\ \text{Inf}(C)\ \&\ \text{Cof}(B)\ \&\ \text{Cof}(C)] \vee [\text{Inf}(B)\ \&\ \text{Inf}(C)\ \&\ \neg\text{Cof}(B)\ \&\ \neg\text{Cof}(C)])]$

Assuming for simplicity that we are looking only at sets of natural numbers (as I will shortly remark this is available with any such abstraction principle), this principle has two different 'infinite' numbers assigned to infinite sets, namely a and b: it gives the same number a to sets such as the even, the odd, and the prime numbers. It gives b to sets such as N-{0}, N-{1}, N-{7, 12} etc. As every infinite set of natural numbers either has an infinite complement or is co-finite in N, all subsets of natural numbers are assigned a cardinal number. Notice that the part–whole principle fails since {0, 2, 4, 6, 8...} is assigned the same number as its subset {4, 8, 12, 16, ... }.

From BP we can easily generalize to get countably infinite many different new abstraction principles. For each n in the natural numbers, define in pure second-order logic the property 'there are exactly n objects falling under C', abbreviated $S_n(C)$. For each n, the following is an abstraction principle that generalizes BP, for $n \geq 1$:

BP$_n$: $(\forall B)(\forall C)\ [!_n x{:}(Bx) = !_n x{:}(Cx) \leftrightarrow ([\text{Fin}(B)\ \&\ \text{Fin}(C)\ \&\ B \approx C] \vee [\text{Inf}(B)\ \&\ \text{Inf}(C)\ \&\ \text{Cof}(B)\ \&\ S_n(\neg B)\ \&\ \text{Cof}(C)\ \&\ S_n(\neg C)] \vee [\text{Inf}(B)\ \&\ \text{Inf}(C)\ \&\ \text{Cof}(B)\ \&\ \neg S_n(\neg B)\ \&\ \text{Cof}(C)\ \&\ \neg S_n(\neg C)] \vee [\text{Inf}(B)\ \&\ \text{Inf}(C)\ \&\ \neg\text{Cof}(B)\ \&\ \neg\text{Cof}(C)])]$

For each n, BP$_n$ gives a different cardinality to infinite sets depending on whether they are co-finite of size n.[29] Thus for $n = 1$, the cardinality of {1, 2, 3, 4...} is the same as that of {0, 1, 3, 4,...} but different from that of {0, 1, 4, 5,...}. Notice that BP$_1$ assigns the same cardinal number to infinite concepts of cofinality $\neq 1$. For each choice of n, a model for this system will assign an object a to all the infinite

---

[29] There are several variations that could be looked at. For instance we could assign different numbers for each class of infinite sets of co-finality $1, 2, \ldots, n$ and then the same number to all infinite sets of finite cofinality $> n$. Or one could assign the same number to all the infinite co-finite sets whose complement has size $\leq n$ and a single number to all the infinite co-finite sets whose complement is $> n$.

co-finite sets whose complement has cardinality different from n, an object b to all the infinite co-finite sets whose complement has cardinality equal to n, and one object c for all those infinite sets whose complement is infinite. And, of course, it will assign a cardinality to every finite set in the standard way.

Notice that PP defeats all infinite instances of part–whole, for if A and B are infinite, with A strictly included in B, then $\clubsuit(A) = \clubsuit(B)$. By assigning the same number to all infinite sets, PP cannot respect part–whole relations between infinite concepts. That's not the case with BP, however. Indeed, the even numbers are classed in the equivalence class of infinite sets with infinite complement. But N-{1} which contains all the even numbers is classed in a different class of equivalence. But of course, there are plenty of instances in which sets and proper subsets are assigned the same object by BP, for instance the even numbers and the multiples of four.

Another principle that satisfies some infinite instances of part–whole is the following co-finite principle.

CF: $(\forall B)(\forall C) [©x:(Bx) = ©x:(Cx) \leftrightarrow ([\text{Fin}(B) \& \text{Fin}(C) \& B \approx C] \vee [\text{Inf}(B) \& \text{Inf}(C) \& \text{Cof}(B) \& \text{Cof}(C) \& \neg B \approx \neg C] \vee [\text{Inf}(B) \& \text{Inf}(C) \& \neg\text{Cof}(B) \& \neg\text{Cof}(C)])]$

This yields mirror images of the cardinals of finite sets (for the co-finite sets) and collapses all infinite sets that have infinite complement to the same cardinality.[30]

Having mentioned Bolzano's interest in preserving the part-whole principle and the fact that some of these abstraction principles preserve part–whole in some, but not all, instances, this might be a good place to mention that no non-inflationary abstraction principle for an operator $ can yield a part–whole theory of 'size' (whether it be called cardinality, numerosity or other) based on $, since no such abstraction principle allows us to prove without exceptions that $A \subset B \rightarrow \$(A) \neq \$(B)$.[31] Suppose we had such an abstraction principle given by an equivalence relation $\approx_\$$. Then the principle would look as follows:

$$\$(A) = \$(B) \text{ iff } A \approx_\$ B$$

[30] In email correspondence, Roy Cook has pointed out the similarity between CP and another principle he calls BCA:

$(\forall B)(\forall C) [\beta x:(Bx) = \beta x:(Cx) \leftrightarrow (B \approx C \& \neg B \approx \neg C)]$

This is essentially the principle of bi-cardinalities stated in Fine (2002, p. 138). On countably-infinite domains, BCA is equivalent to what one obtains by replacing in CF finite and cofinite with 'big' and 'co-big', respectively. Cook also raised the interesting problem of which of the two principles comes closer to adhering to part–whole intuitions.

[31] Indeed, a part–whole theory of 'size' (whether we call it cardinality, numerosity, or any other name is indifferent here) should satisfy the principle:

$A \subset B \rightarrow \text{Size}(A) < \text{Size}(B)$ from which it should be inferrable that: $A \subset B \rightarrow \text{Size}(A) \neq \text{Size}(B)$.

I now observe that such a principle is rampantly inflationary (see Cook 2002), i.e. every $ function satisfying the above constraints in any model whatsoever requires more equivalence classes than there are objects. The observation rests on a proof of a result that is, as Kanamori (1997) has shown, *essentially* already contained in one part of the 1904 Zermelo's proof of the equivalence between the axiom of choice and the well-ordering principle. The part in question (which does not use the axiom of choice) can be seen as establishing that for any set X and *any* function $f: \text{Pow}(X) \rightarrow X$ there are always sets A and B such that A is properly included in B and $f(A) = f(B)$.[32]

This means that any theory that generalizes the part–whole principle for 'size' assignments, cannot be the consequence of an abstraction principle in HP style, unless the neo-logicist is willing to postulate from the outset that the universe is a proper class (or, more weakly, that the universe has no cardinality). If one eschews commitment to a proper class, then the abstraction principles SP, considered in appendix 2, cannot be acceptable to the neo-logicist, for it would preserve the part–whole principle in its entirety. This does not show that the neo-logicist cannot develop theories for which the part–whole principle holds. Indeed, since he can develop (by means of HP) the theory of natural numbers, for which the part–whole principle holds, he obviously can. What cannot be done is to develop a theory that validates the part–whole principle for all concepts. On the other hand, there is at the moment no theory of numerosities for the entire set-theoretic universe.

In the next section we will see that the abstraction principles discussed in this section (PP, BP and its infinite variants, and CF) are all quite powerful, for they can all be used to prove the axioms of PA_2. In addition, I will extend the set of abstraction principles related to HP to include a countable infinity of principles suggested by Richard Heck.

---

[32] I thank Akihiro Kanamori for having pointed out to me the relevance of Zermelo's proof in this connection. The result cited in the text is spelled out in Corollary 2.3 in Kanamori (1997). I take advantage of this mention of Zermelo's theorem on the equivalence between the axiom of choice and the well-ordering principle to mention another important matter. Issues about the axiom of choice are implicit in many of the developments treated in this chapter or their vicinity. For instance, Boolos's proof that HP is satisfied on any infinite domain uses choice to well-order the domain. The use is essential as there is a model of ZF on which HP cannot be satisfied on the continuum. Similar results hold for full ZFC and HP considered on the full set-theoretic universe. I will not pursue these important issues in this chapter but I wanted to flag their importance. I am grateful to Stewart Shapiro for conversations on this and other connected matters.

4.5 Neo-logicism and Finite Hume's Principle

The discussion about Hume's Principle has ignored most of the historical evidence described above, although very recently Richard Heck (2011) referred to the alternative conceptions for numbering infinite sets as a source of further arguments against accepting Hume's Principle as analytic. Let me explain in which context this took place.

In the late 1990s, Boolos began the investigation of the relative strengths of two types of system for second-order arithmetic. The first is based on the standard Dedekind-Peano axioms and characterizes the numbers by means of an ordinal sequence. The second is based on Frege's second-order systems with concepts and objects and captures the notion of natural number as cardinal number. The main result of those investigations can be summarized as follows:

1) the two systems are equi-interpretable and equi-consistent;
2) Frege's system with HP is stronger, in a sense to be explained, than PA_2.

In 1997, Richard Heck developed these investigations by focusing on five systems, one of which restricted HP to finite concepts:
HPF (Finite Hume's Principle):

$$(\forall B)(\forall C)[(\text{Fin}(B) \vee \text{Fin}(C)) \rightarrow (\#x{:}(Bx) = \#x{:}(Cx) \leftrightarrow B \approx C)]$$

Since, as we have seen, $\text{Fin}(B)$ can be expressed in pure second-order logic in a number of ways, this statement can be expressed in the pure language of second-order logic plus the operator #. The important thing about this principle is that it states that finite concepts have the same number if and only if there is a one–one correlation between their extensions and that the number of an infinite concept is different from that of a finite concept. However, it leaves the conditions for when two infinite concepts have the same number completely undecided. This despite the fact that for any infinite concept B we do have a denoting term $\#x{:}(Bx)$.[33] The theory obtained by adding HPF to a standard version of second-order logic is called FAF (Finite Frege Arithmetic). Heck (1997*a*) proved that from FAF one can derive the axioms of second-order Peano arithmetic (PA_2).

A weaker principle than HPF is Weak Hume's Principle (WHP):

$$(\forall B)(\forall C)[(\text{Fin}(B) \,\&\, \text{Fin}(C)) \rightarrow (\#x{:}(Bx) = \#x{:}(Cx) \leftrightarrow B \approx C)]$$

---

[33] As I have already observed, there are ways to get around the totality of the # operator but having a total operator is more in line with Frege and in practice it is easier to work with.

Despite the superficial similarity with HPF, WHP is extremely weak.[34] However, Heck showed (personal correspondence) that if an abstraction principle satisfies WHP and it also satisfies the additional condition 'if Fin(B) & $\neg$ Fin(C) then #x:(Bx) $\neq$ #x:(Cx)', then that abstraction principle satisfies HPF. The key to the proof is showing that if Fin(B) & $\neg$ Fin(C) then there is no one to one and onto mapping between B and C. All the abstraction principles I presented in the previous section satisfy WHP and the additional condition; thus, they satisfy HPF. And since the axioms of PA_2 are derivable from FAF, this guarantees that from any of the abstraction principles I considered one can prove the axioms of PA_2. This technical result will be key to the 'good company' objection to be stated below.

Heck developed a way (using bridge theories) to compare various systems of second order arithmetic and Frege systems according to a suitable proof-theoretic notion of 'stronger than' ($\Rightarrow$) and showed that according to this notion[35] Frege Arithmetic (FA, i.e. second-order logic plus HP) is strictly stronger then Finite Frege Arithmetic (FAF, i.e. second-order logic with HPF as only axiom) which, in turn, is strictly stronger than second-order arithmetic (the Peano version, PA_2):

$$\text{FA} \Rightarrow \text{PAS} \Rightarrow (\text{PAF} \Leftrightarrow \text{FAF}) \Rightarrow PA_2.^{36}$$

In section 2 of his 1997a paper (or, equivalently, section 11.2 of Heck 2011), Heck had emphasized the major philosophical upshots of the investigation. He offered an argument meant to establish that even granting the analyticity of HP (or its alleged status as conceptual truth) one could not thereby conclude to the analyticity of the truths of arithmetic "as we ordinarily understand them [...] but only that arithmetic can be interpreted in some analytically true theory" (Heck 2011, pp. 244–245). The argument, aimed at Wright, used the strength of FA (in comparison to FAF and PA_2) to notice that "the additional proof-theoretic strength of FA, as compared with PA_2, reflects a very real, and very large, conceptual gap between second-order arithmetic and the general theory of cardinality." The extent of the conceptual gap was then spelled out in a twofold way. First, the extension of the equinumerosity criterion for determining the cardinality of infinite sets "constituted as enormous a *conceptual* advance as his [Cantor's] introduction of transfinite numbers was a mathematical advance" (Heck 2011,

---

[34] For instance, it is consistent with WHP that all infinite concepts have size 6.

[35] I will not say more about how this notion of 'stronger than' is defined. This would require too many details that are not essential for an understanding of what follows.

[36] PAF is like PA_2 but with the stronger condition that the predecessor of a number is a number (this captures "predecessors are finite"); in PA_2 we only know that there is a predecessor for each number but the predecessor is not necessarily a number. PAS stands for Strong Peano Arithmetic; it strengthens PA_2 by requesting that nothing (not only no natural number) is a predecessor of 0. For technical details see Heck (1997a) and (2011).

p. 244). But one could object that even though this conceptual advance was obtained very late, this should not block the role of HP as a principle of arithmetic. To this possible objection, Heck retorted that the epistemological aims of the logicist project could only be satisfied by showing that the truths of arithmetic, as we ordinarily understand them, are analytic. Heck's point was that it is not enough to start from any given conceptual truth (say HP, if we grant HP this status) to infer what look like axioms of arithmetic. If this were all that was needed the interpretability of geometry in analysis (granting analysis to be analytic) would show geometry to be analytic (contrary to Frege's and most philosophers' belief). What is needed, according to Heck, is the uncovering of basic laws of arithmetic by reflection on ordinary arithmetical reasoning and from which the ordinary laws of arithmetic could be derived.

If Frege's Theorem is to have the kind of interest Wright suggests, it must be possible to rec-ognize the truth of HP by reflecting on fundamental features of arithmetical reasoning—by which I mean reasoning about, and with, finite numbers, since the epistemological status of arithmetic is what is at issue. For what the logicist must establish is something like this: That there is, implicit in the most basic features of arithmetical thought, a commitment to certain principles, the (tacit) recognition of whose truth is a necessary precondition of arithmetical reasoning, and from which all axioms of arithmetic follow. Having identified these basic laws, we will then be in a position to discuss the question whether they are analytic, or conceptual truths, or what have you. (Heck 2011, p. 245)

Since PA$_2$ is normally taken to include all the principles that can be obtained by reflection on 'ordinary arithmetical thought', then FA cannot be obtained by such reflection (assuming the outcome of such reflection can be written down in the form of a proof) and its content outstrips arithmetical content. Heck identified the locus of this extra content in the unrestricted HP and claimed that HPF was not subject to the same objections. No amount of ordinary arithmetical reflection, he argued, could lead us to assert that the number of the natural numbers is the same as that of the even numbers. Moreover, it would seem very odd to base our knowledge of arithmetic on a concept, such as that of our present concept of cardinality, which did not emerge until the late nineteenth century. And this even if it turns out that HP is analytic of our present concept of cardinality. He concluded:

The objection is not that HP is not known by ordinary speakers, nor that there was a time when the truths of arithmetic were known but HP was not. It is that, even if HP is thought of as 'defining' or 'introducing' or 'explaining' our present concept of cardinality, the con-ceptual resources required if one is so much as to recognize the coherence of this concept (let alone HP's truth) vastly outstrip the conceptual resources employed in arithmetical reasoning. Wright's version of logicism is therefore untenable. (Heck, 2011, p. 246)

Of course, this objection also applies to any abstraction principle that yields PAS or at least, just like all the abstraction principles I introduced, FAF (which outstrips PA_2).[37]

As I already mentioned, it was Heck's claim that HPF does not suffer from the same problems. First of all, recognizing the truth of HPF does not require the conceptual leap made by Cantor. Secondly, one can be convinced of the truth of HPF by reflection on ordinary arithmetical thought. On the first point, Heck noted that HPF was generally accepted by everyone well before Cantor and cites Bolzano as an example of someone who rejected full HP but accepted HPF. Heck then provided an argument, based on the notion of finite counting (an enumeration

---

[37] Although adding PP (the principle that assigns the same number to all infinite sets) to second-order logic allows us to derive the axioms of PAS, one needs to be careful when arguing that an abstraction principle allows the derivation of PAS, as opposed to the weaker FAF. In personal email communication, Richard Heck has given the following characterization based on his paper "Cardinality, Counting, and Equinumerosity" (Heck 2000) and which he traces back to Boolos (1996). I quote the definition and the theorem from his email verbatim (*caveat lector*: Heck uses Nx:Fx in place of #x:(Fx)).

Definition: Say that $\sim$ is *a notion of equinumerosity* iff:

(ZCE) $F \sim [x:x{\neq}x]$ iff $\neg\exists x(Fx)$
(APC) $Nx{:}Fx = Nx{:}Gx \ \& \ \neg Fa \ \& \ \neg Gb \rightarrow Nx{:}(Fx \vee x = a) = Nx{:}(Gx \vee x = b)$
(RPC) $Nx{:}Fx = Nx{:}Gx \ \& \ Fa \ \& \ Gb \rightarrow Nx{:}(Fx \ \& \ x \neq a) = Nx{:}(Gx \ \& \ x \neq b)$

So, in short: No concept other than an empty one is equinumerous with an empty one; if two concepts are equinumerous, then adjoining a new object to each of them preserves equinumerosity, and so does removing an object from each of them.

Theorem: Suppose $\sim$ is a second-order definable notion of equinumerosity. Let HP$\sim$ be:

$$\#F = \#G \leftrightarrow F \sim G$$

Then second-order logic plus HP$\sim$ plus Frege's definitions (for #) implies the axioms of PAS (for #).

Proof: ZCE, APC and RPC are exactly what we need for the proofs of the obviously corresponding axioms of PAS. Three of the others follow from Frege's definitions, given enough comprehension (and we are assuming unrestricted comprehension, so of course we have enough). The existence of successors (for naturals!) then follows from the other axioms, as Boolos observed. QED.

Heck also observed that Boolos proved the converse: If we have PAS for #, and we also have HP$\sim$ for #, then $\sim$ is a notion of equinumerosity. Thus PP is based upon a notion of equinumerosity, but the BP_n, in general, are not.

In the same personal communication, which was a reply to a message of mine discussing the good companions of the previous section, Heck proved that there are countably-infinite many, purely logical notions of equinumerosity in the sense given above. Hence, we have countably-infinite many abstraction principles that prove the axioms of PAS. The abstraction principles in question are of the following form. Let κ be an infinite cardinal. Say that F and G are κ-variants, written $V\kappa(F, G)$ if the cardinality of the symmetric difference of F and G is $\leq \kappa$. Define the following equivalence relation:

$$F \sim_\kappa G \leftrightarrow (\text{Fin}(F) \ \& \ \text{Fin}(G) \ \& \ F \approx G) \vee (\text{Inf}(F) \ \& \ \text{Inf}(G) \ \& \ V\kappa(F, G)).$$

It is easy to show that if $\kappa \neq \mu$ are infinite cardinals then the abstraction relations that are purely logically definable $\sim \kappa$ and $\sim \mu$ are distinct (in that they have different models) and thus we have two different abstraction principles associated with them. Finally, for each n, one can express in second-order logic that a concept has cardinality $\leq \aleph_n$ and that gives us the countably many equivalence relations needed for countably-infinite many abstraction principles.

by means of some process that eventually terminates), aimed at showing how one could have arrived at the truth of HPF using ordinary arithmetical reasoning. The crucial step here would be the realization that "the process of counting, which lies at the root of our assignment of numbers to finite concepts, already involves the notion of a one–one correspondence". I will not rehearse the argument, for establishing the truth or analyticity of HPF will not be my main focus here and for the same reason I will not delve into the problem of the impredicativity of HPF or that related to bad company objections (see Linnebo, 2009a, a special issue of *Synthese* on bad company objections). However, as I will have to refer to bad company objections along the way, I have added an appendix (appendix 1) which briefly explains what bad company objections are and what consequences they have for determining the class of acceptable abstraction principles.

Heck's paper was criticized by MacBride in a paper titled "On Finite Hume" (2000). Contrary to Heck, MacBride claimed that a subject "who has mastered second-order logic but has yet to be introduced to any characteristically mathematical notions" could arrive a priori at the axioms of arithmetic by means of the stipulation of HP, whose formulation requires only logical notions. Against the objection raised by Heck that so far we would have only a modeling of arithmetical truths in FA but not yet a proof that those truths are arithmetical in character, which could only be obtained if HP is arithmetical in character, MacBride retorted that this request was unwarranted. MacBride's arguments can be summarized as follows:

1) It is contentious to claim that prior to the reception of Cantor's work it was finite Hume rather than the full Hume Principle that informed arithmetical thinking;

2) It should be questioned whether finite Hume informs arithmetical reasoning. This attack is two-pronged. If there is an abstraction principle underlying ordinary arithmetic, it is WHP rather than HPF. Second, it is questionable that there is "any abstraction principle even resembling finite Hume that is implicit in ordinary arithmetic practice".

3) One should question Heck's assumption that an arithmetical epistemology must be derived from principles that we can retrieve by reflection on ordinary arithmetic reasoning.

4) Heck's objections reveal a serious misunderstanding of the neo-logicist project, which is not the hermeneutic one of showing that ordinary arithmetical reasoning is a priori.

Heck has replied to all these charges in the recent appendix added to the reprint of his 1997a article in his book *Frege's Theorem* (2011). In the appendix, Heck made explicit the two important issues that had emerged in neo-logicism and had constituted the background of his paper on Finite Hume's Principle, namely the problem about surplus content and the problem about anti-zero (the number of all objects). However, much of the discussion that follows can bypass those specific two issues and rely on Heck's positive proposal:

A clearer statement of it [Heck's positive proposal] emerges towards the end of Chapter 7 and rests upon a distinction, on which Dummett (1988) rightly insists, between Frege's analysis of "just as many" and the abstraction principle for cardinal numbers. The abstraction principle is really just:

$Nx:Fx = Nx:Gx$ iff there are just as many Fs as Gs

And HP in its usual form results from incorporating into this principle Frege's analysis of "just as many" in terms of equinumerosity. My suggestion, then, is that the ordinary notion just as many cannot be so analyzed, because it does not commit itself when both F and G are infinite. The only thing you can get out of the ordinary concept just as many is that there are just as many Fs as Gs if they are equinumerous and at least one of them is finite. Putting that together with the abstraction principle in its pure form then gives us what I called HPF, above. (Heck 2011, p. 263)

Heck's positive proposal consists in parsing "just as many" through what he calls its practical correlate "just enough". And here is how we reach the limits of one–one correspondence when we get to infinite sets:

For there to be just as many cookies as children means that, "... if you start giving cookies to children, and you're careful not to give another cookie to anyone to whom you've already given one, then you won't run out, though you won't have any left". But isn't it implicit in this that "just as many" should be analyzed in terms of one–one correspondence? Well, is it true that, if you start giving even numbers to natural numbers you won't run out but won't have any left? Clearly not. It seems as if "just enough", as we ordinarily understand it, fails to get any grip here. (Heck 2011, p. 263)

Heck's conclusion, which has strong affinities to Schröder's approach described previously, can be reached even without parsing "just as many" through the practical correlate "just enough". As Heck himself recognizes, the Cantorian solution of identifying the cardinal number of the natural numbers and of the even numbers on account of the existence of a one–one correspondence required a major conceptual leap and imposed itself not because it gave an explication of "just as many" that did justice to *all* the pre-theoretical intuitions but rather despite its failure at doing so. The problem here is of course that, as we have already seen, pre-theoretical intuitions seem to be pulling in different directions when confronted with the need to generalize the notion of (cardinal) number to infinite sets.

Thus, Heck ends up claiming that HPF is what underlies our arithmetical thought. As we have seen, in "On Finite Hume" (2000), MacBride challenged this, by discussing various aspects of Galileo's paradox. It would be interesting and tempting to pursue an evaluation of the Heck–MacBride debate, for it involves historical and systematic issues related to Galileo's paradox, the comparison between part–whole and one–one correlation principles, etc. However, in the interest of brevity I will resist that temptation and go straight to what I take to be the core of their disagreement. I am referring to the criticism raised by MacBride in points 3 and 4 listed above.

MacBride claims that Heck "makes no attempt to justify the assumption that an arithmetical epistemology must be derived from principles that we can retrieve by reflection on ordinary arithmetical reasoning". The charge is unwarranted in that Heck does discuss the issue but it does underscore the need for scrutinizing what that argument amounts to. As reconstructed by MacBride, Heck's argument goes as follows:

It is constitutive of arithmetical truths that they are derived from the basic laws that actually inform ordinary arithmetic. We might call these laws the 'canonical sources' of arithmetic truths. Since Hume's principle is not a canonical source, the a priori truths that might be derived from it cannot be arithmetical in character. (MacBride 2000, p. 158)

MacBride finds the argument unpersuasive because "the premise that an arithmetical truth can only be derived from canonical sources is unmotivated". Finally, in the last section of the paper, he diagnosed the source of Heck's worries in the unwarranted assumption that the neo-Fregean project is an hermeneutic one, namely "to show that what we 'had in mind' all along, when we reasoned arithmetically, is *a priori*." Not only is this wrong for Frege,[38] he says, it is also wrong for neo-logicism.[39] The neo-Fregeans do not put forward a thesis about "the ordinary senses of arithmetical expressions". In conclusion: "The objections that Heck (and Benacerraf) voice are ineffective because they misconstrue the nature of the neo-Fregean project. That project never wanted to uncover a priori truth in what we ordinarily think, but to demonstrate how a priori truth could flow from a logical reconstruction of arithmetic." (MacBride 2000, p. 158) In a later paper aimed at Black 2000,[40] MacBride described the opposition—also with reference to Heck's objections—quite vividly:

---

[38] MacBride appeals to some of the secondary literature on Frege, including Dummett (1991) to support the claim. However, the section in Dummett (1991) referred to by MacBride does not seem to me to support MacBride's interpretation. On the question of the hermeneuticity of the Fregean project see Demopoulos (2003) and Blanchette (2012).

[39] However, even in the case of neo-logicism there are conflicting stands concerning the hermeneuticity of the project.

[40] Black's concerns centered on the claim that the number zero is central to neo-logicism but does not seem to have played much of a role in the actual development of arithmetic.

[The] epistemological success of the neo-Fregean programme need not rely upon the effectiveness of (HP) as an "analysis" of ordinary arithmetical notions. Of course, neo-Fregeans do sometimes speak of (HP) as an "analysis" of the ordinary notion of number or "analytic of" that concept (see, for example, Wright 1983, pp. 106–107 and Hale 1997, p. 99). Nevertheless, an alternative epistemology may be gleaned (and extrapolated) from what the neo-Fregeans have to say that makes no relevant play with the notion of analysis and obviates Black's criticism.

Black's criticisms fail to take proper account of the *modal* character of this epistemology. Neo-Fregean epistemology (so envisaged) offers an account of how it is *possible* to acquire knowledge of the fundamental laws of arithmetic (by deriving them from (HP)). It thereby undertakes to describe "an a priori route" (that goes through the recognition of zero) to knowledge of the laws of arithmetic (Wright 1997, pp. 279–80). But it is not thereby committed, as Black assumes, to saying this route is "the" only one available. It is consistent with our coming to recognize arithmetical truths in one way that we could have, and perhaps do, come to recognize their truths by different means. (MacBride 2002, p. 99 [page numbers from the reprint in Cook 2007])

And, once again, MacBride opposed the reconstructive character of neo-Fregean epistemology by contrasting it with the different project requesting that "(HP) can only discharge a foundational role if it makes explicit the principles that actually underlie established arithmetical usage." (MacBride 2002, p. 99) At the same time, MacBride was aware (see MacBride 2002, p. 100) that even granting to the neo-logicist the apriori nature of the 'arithmetical' laws derived in Frege Arithmetic, an additional argument would still be needed to transfer that aprioricity to the laws of ordinary arithmetic.[41] MacBride described again two possible ways to go for the neo-logicist, the reconstructive and the hermeneutic.

From MacBride's replies to Heck and Black we thus see that different interpretations of what the exact nature of the epistemological project of neo-logicism amounts to will have interesting consequences regarding the issue of the acceptability of a reconstruction of arithmetic that deviates from the route that was effectively—even if tacitly—followed in acquiring a grasp of its laws and the issue of whether there might not be several "a priori routes" to knowledge of arithmetic. Notice that the good companions of HP discussed in the previous section have already provided us with specific instances of such alternative a priori routes to the natural numbers and as such they will help us appreciate more clearly what is at stake in the debate.

The MacBride–Heck debate contains several important aspects that also highlight the importance of how we interpret the historical developments but the central one for what follows in this chapter concerns the opposition between the

---

[41] See also Demopoulos (2003) for an insightful discussion of this 'modeling' problem.

HUME'S PRINCIPLE AND THE ASSIGNMENT OF NUMBERS   183

hermeneutic and the reconstructive understanding of neo-logicism. I will come back to it after I discuss the 'good company' objection to be presented next.

4.6 The 'good company' objection as a generalization of Heck's argument[42]

We have seen that, taking his start from the theory of numerosities, Heck offered the following argument against the status of HP as conceptual truth:

But if it is conceptually possible that infinite cardinals do not obey HP, then it is conceptually possible that HP is false, which means that HP is not a conceptual truth, so HP is not implicit in ordinary arithmetical thought. (Heck 2011, p. 266)

In order to evaluate Heck's claim, then, we first need to be clear about the technical situation. Let me recall some of what I have already said about Cantorian cardinalities and numerosities. What I will say is limited to countable sets but the limitation is only for ease of exposition and it is not essential. Consider the natural numbers and all the subsets of the natural numbers. From the mathematical point of view the situation can be described as follows. Cantor introduced an equivalence relation, say $\approx_C$, such that every two countably-infinite sets have the same cardinality, namely $\aleph_0$. In the theory of numerosities developed by Benci, Di Nasso, and Forti, we have a finer assignment of numerosities to infinite subsets of the natural numbers (relative to a labeling of the sets, which I will here take to be identity, and relative to a Ramsey ultrafilter; see Chapter 3 for details). The assignment of numerosities will depend on the particular Ramsey ultrafilter chosen for the construction.[43] But having fixed the ultrafilter we get an associated equivalence relation $\approx_{PW}$. An abstraction principle would then give:

NUM: $NUM(A) = NUM(B)$ iff $A \approx_{PW} B$.[44]

---

[42] In light of the development in the remaining part of this section, I have no qualms in abandoning the word 'objection' in favor of 'consideration' or some other term. The reason why I keep 'objection' is that the argument that led me to the good company objection, namely Heck's argument based on the theory of numerosities, was intended as an objection to the analyticity (or status as conceptual truth) of HP.

[43] Let me also point out that a good company objection, similar to the one I will offer momentarily, could be run on NUM itself, which is defined below. Depending on the choice of a Ramsey ultrafilter (there are uncountably many such choices) the associated NUM numerosities will be in conflict on the assignment of numerosities to infinite sets (while agreeing on numerosities for finite sets and preserving the part–whole principle). If these alternative and non-compatible versions of NUM were taken to capture the same informal notion of numerosity (or cardinality) underlying our ordinary arithmetical reasoning they would be in conflict. Not so if we take them to simply stand next to each other as different ways of classifying infinite collections of numbers according to different equivalence relations.

[44] Of course, in the theory of numerosities this is not postulated as an abstraction principle but rather demonstrated by constructing the numerosities to be the appropriate class of objects.

Since $\approx_C$ and $\approx_{PW}$ are two different relations there is no conflict between them. Of course, the incredible richness of the theory developed by Vieri *et al.* does not stop at the determination of identity conditions (something which in some cases, as in the case of the 'good company' abstraction principles I listed, is very easy to obtain) but in the development of a full algebraic theory of numerosities with a total ordering on them. The two relations $\approx_{PW}$ and $\approx_C$ are not in contradiction. Indeed, $\approx_{PW}$ can be thought of as a refinement of $\approx_C$.

Recall that a bad company objection (see Linnebo, 2009*a, b*) shows that the form of HP that would seem to make it analytic and a priori is shared by other abstraction principles which turn out to be either 1) inconsistent (Basic Law V etc.) or 2) incompatible with HP, for they can only be true on a domain of different cardinality than the one HP requires (in the easiest cases the parity principle (Boolos 1990) and the nuisance principle (Wright 1997) require the universe to be finite whereas HP proves the existence of infinitely many objects); 3) have other defects that have been diagnosed as 'arrogance', lack of modesty, etc. The taxonomy is not exhaustive as one should add, among other forms of bad company, Weir's distraction principles (see Weir 2004) but it is only meant to remind the reader that no amount of effort has been spared to come up with conditions that would separate the 'good' principles from the 'bad' ones (for a rich recent treatment of such matters I refer the reader to Cook 2012; see appendix 1 for a brief introduction). But none of those considerations apply to HP's good companions.

Let us consider, then, the following question. Is Heck's original argument, based on the theories of numerosities, a case of a bad company objection? Yes and No.

Why not? First of all, there is no claim that the conceptual status of the abstraction based on numerosities is on a par with that given by HP. In fact, to specify the equivalence relation for numerosities one needs to appeal to the existence of a Ramsey ultrafilter, i.e. a special non-principal ultrafilter on N. The appeal to N does not make the definition non-logical yet (one could appeal to Frege's logicist construction of N—granting its logicality for the sake of the argument—using HPF or HP) but the existence of a Ramsey ultrafilter turns out to be independent of ZFC, thus independent of the resources of pure second-order logic even enriched with HPF or HP (or any other of the good company abstractions, if one wanted to appeal to them to generate N). Moreover, we have seen that the two equivalence relations are compatible whereas what happens with some typical cases of bad company objections is that two different abstraction principles can only be made true if the universe[45] satisfying one principle has incompatible properties with the

---

[45] HP only requires the universe of objects to be at least countably infinite. PP is the same. By contrast, NUM requires at least uncountably many objects, just as SP (see appendix 2) does.

universe satisfying the other principle.[46] Another form of bad company objection points to the fact that certain abstraction principles, such as Basic Law V, look just like HP but turned out to be inconsistent. However, the equivalence relation on numerosities can be shown to be consistent (in set theory under the assumption that a Ramsey ultrafilter exists; admittedly, FA has a better epistemic validation as it is equi-consistent with PA_2).

Why yes? Under the assumption that both one to one correspondence and the equivalence principle for numerosities capture a principle about the concept of cardinality *underlying our informal arithmetical reasoning* then we are faced with two contradictory principles, for the intuitive notion of cardinality would now be captured both by the neo-logicist # operator (satisfying HP) and by the NUM operator defined by Vieri *et al.* But since they diverge on infinite collections of natural numbers (#(Even) = #(N) but NUM(Even) $\neq$ NUM(N)) they stand in conflict. Notice that the conflict here is of a different nature than the one between HP and HPF. In the latter case, HP is compatible with HPF— since HPF is silent on specific assignments of cardinality to infinite sets—even if we think of both of them as principles underlying the same informal notion of cardinality in ordinary arithmetical reasoning. But that's not the case with HP and NUM.

There is thus a lack of symmetry between HP and NUM. In Heck's argument NUM is used to put pressure on the analyticity of HP but not vice versa, since NUM is not a prima facie candidate for the kind of analyticity that is ascribed to HP on account of the fact that HP's right-hand side of the equivalence is stable in purely logical terms. My generalization of Heck's argument provides infinitely many cases in which symmetry holds and makes the appeal to numerosities dispensable.

Suppose for the moment that Heck's argument is sound. Then, one obvious consequence, by parity of argument, is that not only the abstraction principle(s) connected to the theory of numerosities but also none of the abstraction principles HP, PP, BP and its variants, CF, and the countably infinite many abstractions originating from the equivalence relations $\sim \kappa$ (κ an infinite cardinal of the form $\aleph_n$, for some n) can be conceptual truths (or implicit in ordinary arithmetical thought), for they knock each other out. To each one of them we could apply

---

[46] For instance, the parity principle (Boolos 1990) and the nuisance principle (Wright 1997) can only be true on a domain with finitely many objects and are thus incompatible with HP. This is different from the situation we encounter comparing HP and NUM: NUM requires an uncountable domain for it to be true (if the set of original objects is at least countable) but it is not the case that HP can be true only on a countable domain.

the consideration set forth by Heck: it is conceptually possible that any one of them is false, since for any principle we choose there subsists the conceptual possibility that it might be false. The latter claim is grounded in the fact that different principles assign numbers to infinite concepts in ways that contradict each other. But if Heck's argument is correct we should be able to infer, again by parity of argument, that no one of them can be a conceptual truth. Thus, HP and all its good companions are not conceptual truths. This is then the 'good company' objection.

As a result we seem to be stuck in the following dilemma. On the one hand, HP and every one of the good companions can be argued to be a priori, analytic, and therefore true, using the line of argument used by the neo-logicists for Hume's Principle. Indeed, Hume's Principle shares many properties with the other abstraction principles just listed. They are all consistent, mutually compatible, and in good standing (hence the good company).[47] From each one of them one can derive a suitable version of the axioms for PA_2 (by proving first FAF or PAS). Any argument in favor of the analyticity (or status as conceptual truths) of HP would seem to also apply to any other of them.[48]

On the other hand, neither HP nor any of its good companions can be analytic (or have the status of a conceptual truth), as a consequence of the good company objection.

Is there a way out of this dilemma?

To begin with, is it right to say that the abstraction principles contradict each other when assigning numbers to infinite concepts? Of course there is a whiff of paradox to this claim of contradiction. One could say that the two abstraction principles define different notions of cardinality (say $Card_{HP}$ and $Card_{PP}$) and thus that the contradiction is simply apparent. But if we accept this, would we be willing to treat appeal to HP as replaceable for the purposes of the neo-logicist project with an alternative good companion of HP? After all, within the background of second-order logic, PP also allows for the derivation of PA_2. And the same holds of BP and its variants, CP and all the abstractions originating from the $\sim \kappa$ equivalence relations.

---

[47] All of them satisfy the conditions stated in Cook (2012) but are also non-logical according to most of the criteria discussed in Antonelli (2010).

[48] Let us consider PP and HP. Nothing stops us from applying to the epistemological/semantical status of PP the same considerations that neo-logicists apply to HP. Since for both HP and PP we have a derivation of the axioms of second-order arithmetic (i.e. there are just as many 'Frege's theorems' as there are good companions) each such derivation provides an 'a priori route' to arithmetic. Hence both principles are analytic (or conceptual truths) and a priori.

Before analyzing the impact of the good company[49] objection on neo-logicism, let us examine Heck's argument in more detail.

Heck's conclusion from the above is: "it is conceptually possible that HP is false, which means that HP is not a conceptual truth, so HP is not implicit in ordinary arithmetical thought." I do not think the argument is conclusive, for it seems to me to ride on an ambiguity. When we say that "it is conceptually possible that HP is false" we mean that it might be the case that an alternative principle is underlying our ordinary arithmetical ability. But that claim and "HP is not a conceptual truth" do not have the same meaning.[50] HP could still be a conceptual truth, an 'explanation' of what is 'analytic' or 'constitutive' of the # operator, even if # turns out not to be the 'cardinality' operator underlying our arithmetical ability. And if this is granted the only conclusion we can draw is that "HP *might not* be implicit in ordinary arithmetical thought".

---

[49] After I had written the article on which this chapter is based, I found out that the expression 'good company' in the context of bad-company objections had already been used in Shapiro and Ebert (2009) and in the context of discussion of anti-zero in Clark (2004).

In the case of Shapiro and Ebert, their abstractions, which they claim to be good company to HP, are given by implicit definitions in Hilbert's style that the neo-logicist considers 'arrogant'. Indeed, Hale and Wright (2009) classify the Ebert-Shapiro abstractions as 'bad company'. I do not think the accusation of 'arrogance' can be levied against the abstraction principles I have offered and thus the 'good company' label seems more justified in the present case. I am of course excluding NUM whose right-hand side is not logical and could thus be classified as 'arrogant' if presented as an alternative to HP.

Clark (2004) also uses the expression 'good company' but in a very different context from the one used here. In Clark's case, the good company is the one HP would keep if along with set theory it were not to assign a number to the concept $x = x$. The concern, first raised by Boolos, was that Frege Arithmetic and Set theory disagree in that the first assigns a number to the universe while the latter assigns no cardinal to the class of all sets. The disanalogy with my case is obvious. First of all, my use of 'good company' originates from the 'good companions' of HP and, as far as I can tell, this notion is original with me and it is not what is at stake in Clark's argument. Whereas there is perfect symmetry between HP and its good companions, the comparison between Frege Arithmetic and Set theory could certainly not be meant as a clash between principles that were equally acceptable by neo-logicist lights, for the simple fact that no one, at least in this context, claimed the analyticity of set theory nor that it could be obtained by some logical abstraction principle. It is also too quick to object that Frege Arithmetic and Set theory contradict each other on the assignments of numbers to non set-sized concepts, for set theory is simply silent on non-set sized concepts (would we say that HPF and HP contradict each other on infinite concepts? HPF is simply silent on infinite concepts; and the case for the comparison between Frege Arithmetic and set theory is even more delicate as the first deals with concepts and the latter with sets). Finally, Clark's (or better Boolos's) good company objection is driven by anti-zero, an issue that is completely orthogonal to my concerns (all my principles are total on the domain of concepts and would all be modified in the same way if one wanted to exclude anti-zero). Thus, if any similarity remains, it is simply at the level of putting pressure on HP as an analytic principle by considering issues of cardinal assignments to concepts.

[50] Of course, if one reads 'conceptual truth' so narrowly that it is identified with what can be obtained by reflection on our ordinary arithmetical reasoning, then the conclusion might follow. But then I would claim that the common ground for a comparison with neo-logicism has been lost.

Thus, when Heck says that it is conceptually possible that HP be false, we have to read his claim as: NUM could in principle be the informal notion of cardinality underlying our ordinary arithmetical experience. Since NUM and HP are in conflict on infinite sets, HP cannot also be a principle underlying that ordinary experience. I agree that it is conceptually possible that NUM captures the informal notion of cardinality (this conceptual possibility is of course a very idealized one). This shows that HP might not be the right principle underlying that experience. And we have seen that the same conclusion can be reached, by parity of argument, about all the good companions of HP. However, this is compatible with HP being a conceptual truth about # just as much as PP is a conceptual truth about ♣. We thus seem to have dispelled the dilemma raised before. Thus, it is now necessary to ask whether the good company kept by HP amounts after all to a serious worry for the neo-logicist.

I think the consequences arising from the existence of the good companions will make themselves felt differently according to the stripe of neo-logicism we are dealing with. I will first outline three different positions—liberal, moderate, and conservative neo-logicism—and then make a few additional comments on each of them.

A. The *liberal* neo-logicist might claim: who cares if ordinary arithmetic can be recaptured from many different principles? There was never a claim to the uniqueness of HP in this particular context. What counts is that each such principle is of the appropriate form (abstraction), and each abstraction in good standing that allows us to derive PA_2 will do, thereby yielding a priori knowledge of PA_2. This is obviously MacBride's 'modal' position.

B. The *moderate* neo-logicist retreats to weaker principles: one could simply accept that in light of other considerations (such as the ones adduced by Heck) one should only commit to HPF as being the proper principle for developing neo-Fregean arithmetic. Wright seems to be going this way (at least judging from his paper Wright, forthcoming).

C. The *conservative* neo-logicist: HP is the only correct principle. Then we would need an argument against the good company objection that would separate HP from its good companions.

Considerations on A. Whether or to what extent this position is compatible with the spirit that informs neo-Fregeanism (let alone Frege's position) is questionable. MacBride himself points at the co-existence within neo-logicism of two different epistemological projects (one 'modal' and one 'hermeneutic'). But if one insists on decoupling the analyticity of a principle such as HP from the need that the principle be a principle informing our ordinary arithmetical reasoning (thereby giving up any claim of hermeneuticity), then nothing seems to stand in the way of a neo-logicist who were to accept one of the following alternatives: a) any one of the 'good' abstraction principles different from HP (or even NUM) could in principle be underlying our ordinary arithmetical activity but HP is nonetheless analytic of the concept captured by the # operator and as such is sufficient for the claims of neo-logicism; or b) neither NUM, nor HP nor any of the other 'good' abstraction principles underlies our ordinary arithmetical activity but for the reasons given this does not threaten HP as an analytic principle (or status as conceptual truth), for the latter depends on a stipulation and not on a criterion that would make its status dependent on whether it captures a pre-analytic given concept.

Thus, when, as in Heck's argument or in the good company objection, a claim is made to the effect that it is conceptually possible that HP is false, this ascription of 'falsity' must be referred to whatever principle is underlying our ordinary arithmetical activity. If we give up on that connection to an underlying principle, then there is no conflict between HP and NUM (or any of the other 'good' abstraction principles). They are simply 'analytic' of different and compatible notions (concepts). I conclude that the real core of the matter, under option A, is what the relationship between abstraction principles (or other equivalence principles such as NUM) and informal arithmetical reasoning should be. If that connection is too loose then I do not see why HP should be better off or have any pride of place over PP (or any of the other 'good' abstraction principles). Both allow us to derive the relevant axioms for arithmetic and they can both be stated through an abstraction principle whose right-hand side is purely logical. If the connection must be tighter then I do not see that HP or any other 'good' abstractions can be construed as principles underlying our ordinary arithmetical reasoning; only HPF (or something equivalent to it) seems to satisfy this requirement. Under option A, we are also faced with the full strength of a problem to be discussed in section 7. That is, given that each good companion of HP introduces a different sortal concept, how are we to determine which are *the* natural numbers? The problem in question is the problem of cross-sortal identification of abstracta.

Considerations on B. Heck has claimed that one should keep the connection between ordinary arithmetical principles and abstractions tight and has recommended adopting HPF as a viable such principle, which can be justified by appeal to reflection on our ordinary arithmetical reasoning. Although Heck arrived at the conclusion from below, so to speak, my point of view is from above, given by the number of remarkable alternatives to HP in the assignments of numbers to infinite concepts. The situation here reminds one of what happened with non-Euclidean geometries and Kant's notion of the a priori. When faced with the various alternative non-Euclidean geometries one could insist that a priori intuition was essentially 'Euclidean' or simply retreat to a form of Kantianism in which the form of a priori intuition corresponded to 'absolute geometry' or to projective geometry (as Russell claimed), thereby leaving the choice of the acceptance of the parallel postulate or one of its alternatives outside what was 'justified' by a priori intuition.[51] We are faced with many non-Cantorian assignments of cardinalities all of which agree on the finite and which seem to enjoy the same 'good' properties as HP. Under this construal of the situation it is conceptually possible that HP is false. But that's not a problem for the neo-logicist since the a priori way to PA$_2$ is given by HPF and the latter principle is taken to be the principle underlying our arithmetical knowledge.

Considerations on C. I do not see that the neo-logicist could appeal to the success of Cantor's theory in mathematics, for this success does not depend on the claim that the notion of cardinal (given in terms of one to one correspondence) is the appropriate generalization of finite counting. And no direct argument (from below, so to speak) given so far seems to persuasively lead beyond HPF. Here the neo-logicist has to provide positive reasons for knocking out all the good companions of HP showing that despite their posturing they are bad companions after all. What arguments could he/she appeal to?

The usual list of criteria for good abstractions vs. bad abstractions, elaborated in connection to bad company objections, does not seem very helpful here. Our abstraction principles do not seem to defy any of the usual criteria: consistency, field-conservativity, irenicity, stability, etc. (all the principles I presented satisfy the criteria for 'goodness' provided in Cook 2012; see also appendix 1).

The conservative neo-logicist must do one of two things. Either give an a priori argument for why HP is after all the principle that underlies our ordinary arithmetical experience. Or come up with new criteria to show that HP has pride of place over the other abstraction principles and thus that the 'good abstractions' I have mentioned in this chapter can be recast as bad company objections.

While it is obviously impossible to predict what positive proposals might emerge in this connection, I will mention two natural objections to my challenge to the conservative neo-logicist.[52] The first consists in pointing out that "HP enjoys a naturalness and simplicity that is manifestly not shared by the other principles. All but one of them are far too ad hoc and disjunctive to merit any claim to be analytic of our pre-theoretic concept of cardinality." I do not question that HP appears very natural and simple. But the issue is that of determining what is our

---

[51] Another analogy can be obtained with alternative theories of truth whose common core are the grounded truths but diverge on the ungrounded ones. I thank Hannes Leitgeb for having brought this analogous situation to my attention.

[52] These objections have been raised by an anonymous referee.

pre-theoretic notion of cardinality. And what is being questioned here (just as it has been questioned by Heck and others) is whether our pre-theoretic notion of cardinality has anything to say about the assignment of numbers of infinite sets. To many it appears obvious that according to our pre-theoretic notion of cardinality the number of even numbers is smaller than that of the natural numbers. Thus, I do not see that appealing to simplicity and naturalness is going to help the conservative very much. The liberal neo-logicist could, by contrast, appeal to such criteria in preferring the use of HP as opposed to any of the other good companions but this choice would be dictated by factors other than those that relate to our pre-theoretic notion of cardinality (and would thus not result in a threat to any other liberal neo-logicist who preferred to appeal to an alternative to HP in order to carry out the program).

The disjunctiveness of the principles is obvious since all of them, except HP, rely on the disjunction between finite and infinite concepts (recall that this is the notion of Dedekind-finite which is expressible in second-order logic). But, the objection continues, "Surely, our concept of sameness in cardinality does not presuppose the notion of Dedekind-infinity." There are two important issues here. Concerning disjunctiveness, one must address the issue of which disjunctive principles have a claim to define sortal concepts. One way to address the issue would be to try to define purely formally (i.e. without reference to pre-analytic concepts) which criteria allow us to distinguish which are the 'good' disjunctive principles and which are the 'bad' ones. In doing so, the conservative neo-logicist will need to make sure that the many disjunctive principles that appear in neo-logicist developments of the theory of real numbers and set theory are left unscathed, lest his solution do more damage than good. But if the disjunctiveness issue is motivated by the additional claim given above concerning "our concept of sameness of cardinality" than we are back to square one, for the real issue here is whether we have an informal concept of cardinality that extends to infinite concepts and, if it does, whether it coincides with that captured by HP. I, together with Heck and others, reject this claim and thus a mere assertion on the part of the conservative neo-logicist that "our concept of sameness in cardinality" or our "pre-theoretic concept of cardinality" are captured by HP will amount to no more than begging the question.

There is much more to say but I do hope that the good company objection will lead to a better understanding of the nature of the stripe of neo-logicism that is being defended and of whether appeal to HP, as opposed to any alternative good companion of HP, is after all so central to the neo-logicist program.

I will conclude with an application to metaphysics that follows from the existence of the good companions of HP.

4.7 HP's good companions and the problem of cross-sortal identity

It will be useful in this section to raise a problem that turns out to be an issue for any variety of neo-logicism although I think it is especially pressing for the liberal neo-logicist. The problem is that of cross-sortal identity for abstract objects. It was raised a few times in the neo-logicist literature[53] and most clearly studied in Cook and Ebert (2005). The problem concerns under what circumstances the abstracts obtained using an abstraction principle can be identified with the abstracts obtained from a different abstraction principle. In a way, this problem can be considered to be a more subtle and insidious form, yet an instance, of the infamous 'Caesar problem'. The latter concerns how to decide identities of the form $@(C) = $ Caesar (or any term denoting an object not presented as $@(D)$ for some D). While no one thinks that Caesar is an abstract object, it is not in virtue of the corresponding abstraction principle that we can determine that. The latter only settles questions about identities of the form $@(C) = @(D)$ by appealing to the equivalence relation on the right-hand side of the abstraction principle.[54] And even if one were able to bypass the Caesar problem through metaphysical considerations (because, say, one has principles determining that Caesar is contingent while the objects obtained by abstraction are necessary), the problem of cross-sortal identification for abstract objects would still be there. More formally, the problem is the following. Let us consider, for simplicity, abstraction principles on concepts. Assume we have two abstraction principles yielding:

$$(\forall X)(\forall Y)\,(@_1(X) = @_1(Y) \leftrightarrow R_1(X, Y))$$

and

$$(\forall X)(\forall Y)\,(@_2(X) = @_2(Y) \leftrightarrow R_2(X, Y))$$

Under what conditions, if ever, can we set $@_1(C) = @_2(D)$? or, more specifically, $@_1(C) = @_2(C)$?

Cook and Ebert (2005) describe this problem as the C-R Problem for it emerges in issues of identification of the real numbers with (a subset of the) complex numbers, when both systems of numbers are obtained by abstraction. Given the nature of this chapter, my emphasis will be to focus the discussion on cross-sortal identifications of abstracts such as finite cardinals. There are two major strategies for identification that have been considered in the literature. Given two

---

[53] See Hale and Wright (2001b), Fine (2002), and Cook and Ebert (2005).
[54] On the Caesar problem see, among others, Wright (1983), Heck (1997a), MacBride (2003), Hale and Wright (2001b).

abstraction principles $@_1$ and $@_2$, the first strategy states: $@_1(C) = @_2(D)$ if and only if 'R$_1$' and 'R$_2$' express the same equivalence relation and R$_1(C, D)$ is the case. There are several non-equivalent ways of spelling out what it means for 'R$_1$' and 'R$_2$' to express the same equivalence relation. The most obvious proposal is that 'R$_1$' and 'R$_2$' express the same equivalence relation iff $(\forall X)(\forall Y)$ (R$_1(X, Y) \leftrightarrow$ R$_2(X, Y))$. However, according to Cook and Ebert, who refer to Fine (2002), this strategy won't do because it clashes against an important intuition. The intuition concerns two of the principles I have discussed in this chapter, HP and PP.[55] Cook and Ebert, following Fine (2002), correctly claim that, if the universe is uncountable, the numbers yielded by HP and those yielded by PP are distinct. Thus, using the above condition for settling issues of identities of the abstracts generated by HP and PP leads to an unpalatable consequence. They put it this way: "Intuitively, however, the natural numbers provided by these two principles (if both are acceptable) are identical—the natural numbers are a proper sub-collection of the cardinal numbers"(p. 125).[56] Thus they reject the first strategy

[55] The latter is called FHP in Cook and Ebert; as I have already pointed out, this is misleading since it can be confused with HPF (Finite Hume's Principle). Hence I will stick to PP. In their presentation, both Fine (2002) and Cook and Ebert (2005) start with an abstraction principle limited to finite concepts (something like Finite Hume's Principle) and then extend it, by adding the appropriate clause, to infinite concepts in order to obtain an abstraction principle that assigns the same object to all infinite concepts (thereby obtaining PP). At this point the equivalence relations for HP and PP are different since the first applies uniformly the criterion of one to one correspondence to all concepts while the latter applies one to one correspondence to finite concepts and assigns the same number to all infinite concepts. The two abstraction principles can of course be differentiated semantically in models having at least uncountably many objects, say the real numbers. On such a model the natural numbers and the reals would, according to HP, have two different numbers associated to them; but the same number according to PP. The strength of the counterexample (to using the extensional identity of the equivalence relations in determinations of cross-sortal identities) thus partly rests on the strength of the intuition generated by the comparison between HP and PP. While the two relations are different the abstracts corresponding to the finite cardinals yielded by the two principles should coincide. At any rate, this is the intuition expressed by Fine, Cook and Ebert and with which I sympathize.

[56] Similarly, Fine (2002), discussing Hale and Wright's desire to leave room for the possibility that numbers are classes says:

"The focus of their discussion is accordingly on the question of how two abstracts might be the same even though they are associated, through their respective means of abstraction, with different concepts. As I have indicated, this focus appears to be misplaced; for there seems to be no coherent view that will let both numbers be classes and classes be abstracts [I omit a footnote, P.M.] Our focus, on the other hand, is on the question of how two abstracts might be distinct even though they are associated, through their respective means of abstraction, with the same concepts. What might contribute to their being different?

The obvious answer is the identity criteria themselves. On this view, then, two abstracts will be distinct if they are associated with different means of abstraction. But even if one does not believe in the possibility of numbers being classes, this might still appear to be too strong. Suppose one defines *natural number* using equinumerosity as the criterion of identity, but only in application to finite concepts, and that one defines cardinal number again using equinumerosity as the criterion of

because it yields as a consequence that the abstracts generated by HP and those generated by PP are distinct. According to Cook and Ebert, this does not in itself bar the possibility that a more fine-grained approach to the notion of " 'R$_1$' and 'R$_2$' express the same equivalence relation" might yield a solution to the problem of cross-sortal identities but they do not pursue this possibility. I also will not pursue this strategy further because all I need to take from this discussion is to single out the intuition, which I share with Cook, Ebert and Fine, that the (finite) natural numbers yielded by HP and PP should be identical.

The second strategy for identifying abstracts appeals to the identity of the equivalence classes generated by R$_1$ and R$_2$ or, more precisely, it lets the identity of the abstracts covary with the identity of the equivalence classes. Cook and Ebert (2005) consider three possible options[57] and conclude that none of the options work. I will show that their conclusion with respect to the third option can be blocked.

In order to explain what the third option amounts to we need to consider the concept of a restricted abstraction principle. The idea is to limit the range of concepts which are in the range of the second-order quantifiers appearing in the abstraction principle. Thus, rather than, say,

$$(\forall X)(\forall Y)\,(@_1(X) = @_1(Y) \leftrightarrow R_1(X, Y))$$

we will consider more generally

$$(\forall X_{\Phi(X)})(\forall Y_{\Phi(Y)})\,(@_1(X) = @_1(Y) \leftrightarrow R_1(X, Y))$$

where the higher-order predicate Φ picks out a subdomain of all concepts. Ordinary unrestricted abstraction principles are just an instance of such principles (one lets Φ pick out the entire domain of concepts).

---

identity, but now in application to all concepts. Does one want to say that the natural number 0 and the cardinal number 0 are not the same on account of their criteria of identity not being the same?

Of course, one might insist in the face of such an example that a conceptual criterion of identity should have application to all concepts. But suppose that the proposed criterion of identity for natural number is extended to all concepts by treating it as a universal relation on any two infinite concepts. We then obtain a single infinite number ∞ just as in pre-Cantorian mathematics. Do we still want to say that the respective 0s are not the same? These examples force us to face the possibility that the criteria of identity might be different in a way that is not relevant to the identities of the abstracts in question." (Fine 2002, pp. 48–49).

[57] The first option identifies abstracts with the equivalence classes of concepts. For an abstraction operator @ and concept P, the basic idea of this option is that, @(P) is identified with $\{X: R(P, X)\}$, where R is the equivalence relation used to introduce @ by abstraction. In the second option the truth of identities for abstracts covaries with the identities of the equivalence classes, which is still defined using extensions. Thus in the first and second option the notion of extension is central. This is not the case for the third option. The details are not relevant for what follows.

Fine's proposal for identifying abstracta with equivalence classes is formulated by Cook and Ebert (2005, p. 136) as the principle ECIA$_2$. Their formulation of the principle reads as follows. Given two principles

$$AP@_1: (\forall X_{\Phi(X)})(\forall Y_{\Phi(Y)})\, (@_1(X) = @_1(Y) \leftrightarrow R_1(X, Y))$$

and

$$AP@_2: (\forall X_{\Psi(X)})(\forall Y_{\Psi(Y)})\, (@_2(X) = @_2(Y) \leftrightarrow R_2(X, Y))$$

identities are determined by

$$ECIA_2: (\forall X_{\Phi(X)})(\forall Y_{\Psi(Y)})\, [(@_1(X) = @_2(Y) \leftrightarrow ((\forall Z)\, (R_1(X, Z) \leftrightarrow R_2(Y, Z)))].^{58}$$

However, Cook and Ebert (2005) reject this criterion of cross-sortal identity because it is in conflict with a metaphysical principle they defend, namely the principle of uniform identity.

And it is in this connection that the abstraction principles yielding the good companions of HP can help us sharpen our intuitions on Cook and Ebert's proposal which, they claim, is implicit in Hale and Wright's solution to the Caesar problem (Hale and Wright, 2010b). Let us introduce the *principle of uniform identity* (PUI). Suppose we have two abstraction principles (they might even be intended to apply to a restricted class of concepts but this will not matter for my argument). Cook and Ebert explain the idea behind the principle as follows:

The idea is that if it turns out that there is a concept on the shared domain of application whose @$_1$ abstract is identical to its @$_2$ abstract, then, for any concept in its shared domain, its @$_1$ abstract will be identical to its @$_2$ abstract. (Cook and Ebert 2005, p. 129)

The motivating example behind this principle was the issue of the identification of 0 with the empty extension. Assuming that 0 is identical to the empty extension, the principle should license the identification of all natural numbers with extensions. Formally the principle looks as follows. Given

$$AP@_1: (\forall X_{\Phi(X)})(\forall Y_{\Phi(Y)})\, (@_1(X) = @_1(Y) \leftrightarrow R_1(X, Y))$$

and

$$AP@_2: (\forall X_{\Psi(X)})(\forall Y_{\Psi(Y)})\, (@_2(X) = @_2(Y) \leftrightarrow R_2(X, Y))$$

---

[58] The principle actually paraphrases away all reference to equivalence classes by exploiting the fact that @$_1(P) = @_2(Q)$ is true iff $\{X: R_1(P, X)\} = \{X: R_2(Q, X)\}$ and this latter identity can, using extensionality, be rewritten as $(\forall Y)(R_1(P, Y) \leftrightarrow R_2(Q, Y))$.

the Principle of Uniform Identity says:

(PUI) $(\exists X_{\Phi(X)\ \&\ \Psi(X)})\ (@_1(X) = @_2(X)) \rightarrow (\forall X_{\Phi(X)\ \&\ \Psi(X)})$
$(@_1(X) = @_2(X)))$

But applying this principle to the good companions of HP we get some contradictory consequences. For instance HP and PP have the same domain of application
(all the concepts, i.e. the Ψ and Φ in the principle of uniform identity in this case
impose no restriction). Assuming that they have at least 0 in common, something
we were encouraged to do also by Fine and by Cook and Ebert's considerations
earlier on, then it would follow that for all concepts C, $\#x{:}(C) = \clubsuit x{:}(C)$. Why
is that? Assuming that at least one abstract is in common, say $\#x{:}(x \neq x) =$
$\clubsuit x{:}(x \neq x)$ (=0), as we were led to by previous considerations, then consider a
domain containing N, the natural numbers, as well as R, the real numbers, and
let's assume that Nx and Rx (which might be the identity $x = x$ if we impose
that R be the entire domain) are the concepts characterizing both N and R.
We have $\#x{:}(Nx) = \clubsuit x{:}(Nx)$ and $\#x{:}(Rx) = \clubsuit x{:}(Rx)$ by the principle of uniform
identity. But $\clubsuit x{:}(Rx) = \clubsuit x{:}(Nx)$ according to PP; hence by transitivity of equality
$\#x{:}(Nx) = \#x{:}(Rx)$. Contradiction since $\#x{:}(Nx) \neq \#x{:}(Rx)$. Thus, Cook and Ebert
cannot defend both intuitions at once. Either we give up the intuition that HP and
PP 'share' even a single abstract or the principle of uniform identity leads to a
contradiction when applied to simple cases (such as HP and PP). And this I think
provides the independent consideration requested in section 7 of Cook and Ebert
(2005) for resisting the strategy, based on the appeal to the principle of uniform
identity, against the use of ECIA$_2$.[59]

Of course, it is not my intention here to propose a solution to the problem
of cross-sortal identification for abstracts but only to point out that some of the
proposals in the literature might benefit from being discussed by keeping in mind
the context of the good companions to HP.

4.8 Conclusion

I hope the reader of this chapter will come away with four main important points.
The first, historical, is that the evidence from the nineteenth century shows that
using the criterion of one–one correspondence was not the orthodoxy when
it came to assigning numbers to infinite sets and that a variety of alternatives
were considered and defended. Cantor's one–one correspondence criterion for

---

[59] In email correspondence, Cook stated that he now endorses ECIA$_2$ as a criterion of identity for
cross-sortal abstraction.

assignments of cardinality to sets imposed itself despite the fact that it failed to account for all our pre-theoretical intuitions concerning number assignments to sets. Since HP follows Cantor's approach when it comes to assignments of numbers to infinite concepts, the variety of options that were historically explored and defended supports the worry that HP might not be an adequate 'explication' or 'explanation' of our pre-analytic concept of cardinality.

The second point is logico-mathematical. I have shown that there are countably-infinite many logical abstraction principles that, like HP, allow us to derive the axioms of second-order arithmetic. These infinitely many principles, giving rise to as many 'Frege's theorems', satisfy the same criteria for 'good abstractions' that are displayed by HP.

The third point consists in the extended discussion leading to the good company objection, based on the historical and logico-mathematical situations evidenced by the first two points. Generalizing an argument that Heck raised in the context of the theories of numerosities, I offered an analogous objection that uses the good companions of HP to put pressure on the analyticity (or conceptual status) of HP. And while I was quick to show that the objection can be countered, I hope that I was also able to persuade the reader that the situation evidenced by the presence of the good companions allows us to articulate in a sharper way the different varieties of neo-logicism on offer, and especially the tension between hermeneutic versus reconstructive approaches to neo-logicism. Moreover, I argued that the three different varieties of neo-logicism that I described (liberal, moderate, conservative) will be impacted differently by the existence of HP's good companions.

Finally, the fourth point is more of a collection of partial results ranging from the impossibility to postulate a non-inflationary abstraction principle that preserves part-whole relations among (the objects falling under different) concepts to showing how the developments presented in this chapter might have an impact in other areas, such as the metaphysical issue of cross-sortal identities for abstracta generated by different abstraction principles.

4.9 Appendix 1

Bad company objections and admissible abstraction principles

I have claimed that the good companions of HP discussed in the main text satisfy all the formal properties of Hume's Principle. I quickly mentioned some such properties (consistency, stability, etc.). These properties will be familiar to the reader who has followed the ins and outs of the discussion on neo-logicism but not necessarily to a general reader. This appendix is meant for the latter reader. In order not to multiply definitions and symbolisms without necessity, I will stick to the terminology and conventions followed in Cook (2012).

The discussion will be limited to what are called purely logical abstraction principles, that is principles in which all the terms are logical except for the operator $@_E$ which occurs only as displayed:

$$(\forall \alpha)(\forall \beta)[@_E(\alpha) = @_E(\beta) \leftrightarrow E(\alpha, \beta)]$$

As we know, Hume's Principle (HP) is a purely logical abstraction principle. Unfortunately, so is also Basic Law V, which is obtained from the above schema by letting the quantifiers range over concepts and the equivalence relation be equi-extensionality:

$$\text{BLV } (\forall X)(\forall Y)[\text{ext}(X) = \text{ext}(Y) \leftrightarrow \forall z(Xz \leftrightarrow Yz)]$$

This is the first and most obvious example of bad company for HP. Prima facie, BLV has the same logical form as HP but BLV leads to inconsistency, so why should we accept HP? The answer in this case is that a necessary condition for accepting an abstraction principle is that it be satisfiable. We say that a sentence ϕ is κ-satisfiable if and only if there is a model $\mathcal{M}$ with domain of cardinality κ such that ϕ is true in $\mathcal{M}$.

Let us denote with Cook (2012) an abstraction principle that depends on the equivalence relation E with A_E.

Definition. We say that A_E is **satisfiable** if and only if there is a cardinal κ such that A_E is κ-satisfiable.

Note that HP has models of cardinality $\aleph_0$ and thus it is $\aleph_0$-satisfiable. In fact, assuming the axiom of choice, HP can be shown to be κ-satisfiable for any $\kappa \geq \aleph_0$. By contrast, on account of its inconsistency, BLV is not κ-satisfiable for any cardinal κ. One thus restricts attention to abstraction principles that are satisfiable. While satisfiability is a necessary condition for the acceptance of an abstraction principle, it is not sufficient. An early example, concocted by George Boolos, shows that there are satisfiable principles which can only be true in a model if and only if HP is false on that model. The specific principle presented by Boolos is called the parity principle:

Parity Principle: $(\forall X)(\forall Y)[\text{par}(X) = \text{par}(Y) \leftrightarrow FinEven((X \wedge \neg Y) \vee (Y \wedge \neg Y))]$

where $FinEven(Z)$ is the purely logical second-order predicate stating that the objects falling under Z are finite and even.

The parity principle has the property that it can only be satisfied on models with finite domains. Since HP can only be satisfied on infinite domains, it follows that they cannot both be true on the same model. In order to rescue the pride of place assigned to HP by neo-logicists and to guard it against the insidious form of bad company occasioned by the Parity Principle, several more stringent criteria of acceptance have been introduced with the aim of separating the acceptable abstraction principles from their bad companions. One such condition is called field-conservativity but since stating the condition requires a bit of preparatory work, I will skip the definition of field-conservativity and go straight to a condition that implies it.

Definition. A_E is **unbounded** if and only if for any cardinal γ, there is a $\kappa \geq \gamma$ such that A_E is κ-satisfiable.

HP is unbounded while the parity principle (which can only be true on finite domains) is not. But unboundedness cannot be a sufficient condition for discriminating the acceptable abstractions for there are pairwise incompatible unbounded abstraction principles. A standard counterexample presents two purely logical abstraction principles A_{E1} and A_{E2} such that A_{E1} can only be true on models whose domain is a successor cardinal and A_{E2} can only be true on models whose domain is a limit cardinal. However, as we have seen, HP satisfies a stronger property, which we introduce next.

Definition. A_E is **stable** if and only if there is some cardinal γ such that, for all $\kappa \geq \gamma$, A_E is κ-satisfiable.

HP is stable since for all $\kappa \geq \aleph_0$, HP is κ-satisfiable. Once again, it turns out that stability is not strong enough to single out the class of acceptable abstractions.

The dialectic should now be clear. Several proposals have been made for identifying a property that characterizes the good abstractions and new counterexamples (bad company style) require finding stronger discriminating conditions. It is part of the end-goal to preserve the goodness of HP and thus HP also satisfies the next definition. Much of the discussion involves finding the proper notion of conservativity at play in this situation but the conceptual considerations and the logical details cannot be rehearsed here. Once again, I will skip the precise definition of conservativity involved here (called "strictly logical symmetrically class conservativity" [SLSC-conservativity]) and state a property which implies it (for details see Cook 2012).

Definition. A_E is **strongly stable** if and only if, there is some cardinal γ such that, for any κ, A_E is κ-satisfiable if and only if $\kappa \geq \gamma$.

There are many other notions connected to the ones we have defined, such as irenicity, field-conservativity, etc. but the technical details required to introduce these notions would be more confusing than useful for the novice. But the general picture can be grasped by looking at the following properties characterizing classes of purely logical abstraction principles:

Strongly stable $\Rightarrow$ Irenic $\Rightarrow$ Stable $\Rightarrow$ Unbounded $\Rightarrow$ field-conservative $\Rightarrow$ Satisfiable.

The implications are strict with the possible exception of Irenic $\Rightarrow$ Stable and Unbounded $\Rightarrow$ field-conservative for which it is not known whether Irenic $\Leftrightarrow$ Stable and Unbounded $\Leftrightarrow$ field-conservative. It is however known that (Irenic $\Leftrightarrow$ Stable) if and only if (Unbounded $\Leftrightarrow$ field-conservative).

The important upshot here is the following. The most persuasive proposal for delimiting the acceptable abstraction principles offered so far in the literature is given by Cook (2012). He identifies the class of acceptable abstraction principles with the strongly stable ones. Strongly stable principles are SLSC-conservative, stable, irenic, unbounded, field-conservative, and satisfiable.

HP satisfies strong stability. But so do PP, BP and all its varieties BP_n, CF, and all the abstractions based on the notion of κ-variance discussed in the main text.

4.10 Appendix 2

Two assignments of cardinalities by means of abstractions that satisfy the part–whole principle.

The following abstraction principle assigns distinct numbers to each infinite set. Consider the equivalence relation $(\text{Fin}(B)\ \&\ \text{Fin}(C)\ \&\ B \approx C) \vee (\text{Inf}(B)\ \&\ \text{Inf}(C)\ \&\ B = C)$. Define by abstraction:

BLV-F: $(\forall B)(\forall C)\ [\nabla x{:}(Bx) = \nabla x{:}(Cx) \leftrightarrow [(\text{Fin}(B)\ \&\ \text{Fin}(C)\ \&\ B \approx C) \vee (\text{Inf}(B)\ \&\ \text{Inf}(C)\ \&\ B = C)]]$

BLV-F stands for 'Basic Law V except for Finite'. It satisfies the part–whole principle in the following form: if A is properly included in B then $\nabla x{:}(Ax) \neq \nabla x{:}(Bx)$. Since, as I have shown, every abstraction that satisfies the part–whole principle as stated is rampantly inflationary, this abstraction is not part of the 'good company'. The principle is eerily close to Basic Law V from which it diverges only on finite concepts. Since it assigns different abstracts to different infinite concepts one might also legitimately worry about its consistency. Indeed, the worry is fully warranted as in personal correspondence Cook has shown that this principle is formally inconsistent.

I will now propose a different abstraction principle that uses an equivalence relation that has been studied by Frank Sautter. Define first an equivalent relation $\approx_1$ between concepts as follows. A and B are $\approx_1$ iff there is an injective function from A to B, every injective function from A to B is surjective and every injective function from B to A is surjective. Only pairs of finite concepts can satisfy $\approx_1$. However, $\approx_1$ is not an equivalence relation over concepts since reflexivity fails for infinite sets. Now we define $\approx_2$ by means of $\approx_1$: A and B are in the relation $\approx_2$ iff there is a subset C of A and a subset D of B such that $A - C = B - D$, and $C \approx_1 D$. Sautter proved that $\approx_2$ is an equivalence relation. Informally, $\approx_2$ holds between finite sets when they have the same cardinality and between infinite sets A and B when their intersection $A \cap B$ satisfies: $A\text{-}(A \cap B)$ and $B\text{-}(A \cap B)$ are finite and of the same cardinality. Let us consider some examples using as domain the natural numbers. $\{2, 3, 9\}$ and $\{2, 5, 9\}$ will end up in the same equivalence class since their intersection is $\{2, 9\}$ and $\{3\}$ and $\{5\}$ are the two sets C and D standing in the $\approx_1$ relation. If the two finite sets have empty intersection the result holds by virtue of the two sets satisfying the $\approx_1$ relation (using the empty set as C and D). Consider now some examples of infinite sets. $\approx_2$ partitions the universe in such a way that two infinite sets of natural numbers A, B, are in the $\approx_2$ relation if their intersection $A \cap B$ is infinite and $B\text{-}(A \cap B)$ and $A\text{-}(A \cap B)$ are finite and have the same Dedekind-finite size. For instance, $\{1, 2, 4, 6, 8, \text{etc.}\}$ and $\{3, 2, 4, 6, 8, \text{etc.}\}$ are in the same equivalence class according to $\approx_2$. Even $= \{0, 2, 4, 6, 8, 10, \ldots\}$ and Odd $= \{1, 3, 5, 7 \ldots\}$ are not in the same equivalent class (for their intersection is empty and they fail to stand in the $\approx_1$, and hence $\approx_2$, relation). Same for $N = \{0, 1, 2, 3 \ldots\}$ and N-$\{0\}$. Their intersection is N-$\{0\}$ but the two sets N-$(N\text{-}\{0\})$ $(= \{0\})$ and $(N\text{-}\{0\})$-$(N\text{-}\{0\})$ $(= \emptyset)$ cannot be in the relation $\approx_1$. This shows that there are some instances of subset inclusion for infinite sets that are validated by the $\approx_2$ relation. In fact, the subset principle holds generally. Proof: if A and B are finite and A is a proper subset of B that's easy to show. If A is finite and B is infinite and A is a proper subset of B then their intersection $A \cap B$ can at best be finite. Thus, the two sets $A\text{-}(A \cap B) = \emptyset$ and $B\text{-}(A \cap B)$ cannot satisfy the

$\approx_1$ relation. If both sets are infinite and A is a proper subset of B then whatever the size of B-$(A \cap B)$ it cannot be in the $\approx_1$ relation to A-$(A \cap B)$ for the latter is the empty set and the former is not. We can do abstraction on the $\approx_2$ relation to obtain

SP: $(\forall A)(\forall B)\ [\spadesuit x:(Ax) = \spadesuit x:(Bx) \leftrightarrow A \approx_2 B]$.

Since every abstraction that satisfies the part–whole principle as stated is rampantly inflationary, SP is not part of the 'good company'. SP requires uncountably many objects to be satisfied on a domain such as the natural numbers.[60]

---

[60] A more direct proof for SP that does not use Zermelo's result is as follows. If we look at the classes of infinite sets under the equivalence $\approx_2$ we get uncountably many classes. One can see this by reflecting on the fact that there are at least countably many equivalence classes: each set $\{p, p^2, p^3, \ldots\}$ with p prime gives rise to a different class. Each class must contain only countably many sets. The reason is that if C is the equivalence class of an infinite set under $\approx_2$ we can determine the number of elements in it by looking at a representative A and asking how many possible sets B can stand in the relation $\approx_2$ to it. Well, if B can be in there it is because there are C and D finite satisfying the $\approx_2$ relation. But there can be only countably many such. Thus we have countably many elements in each equivalence class. Now assume there are only countably many classes. If we take the union we end up with countably many subsets of the natural numbers, contrary to the fact that $\approx_2$ is an equivalence relation on the power set of the natural number (and thus of uncountable cardinality).

Bibliography

Acerbi, F., 2009, Transitivity cannot explain perfect syllogism, *Rhizai*, VI, 23–42.

Albertus de Saxonia, 1492, *Questiones Subtilissime in Libros Aristotelis de Celo et Mundo*, Venetiis. Reprint Georg Olms, Hildesheim, 1986.

Angelelli, I., 1984, Frege and abstraction, *Philosophia Naturalis*, 21, 453–471.

Angelelli, I., 2004, Adventures of abstraction, *Poznan Studies in the Philosophy of the Sciences and the Humanities*, 82, 11–35.

Angelelli, I., 2013, Abstracción y pseudo-abstracción en la historia de la lógica, *Notae Philosophicae Scientiae Formalis*, 2, 87–105.

Antonelli, A., 2010, Notions of invariance for abstraction principles, *Philosophia Mathematica*, 18, 276–292.

Argand, R., 1806, *Essai sur une maniére de représenter les quantités géométriques*, Paris. Cited from the second edition published in 1874 by the publisher Gauthier-Villars in Paris.

Arthur, R., 1999, Infinite number and the world soul; in defense of Carlin and Leibniz, *The Leibniz Review*, 9, 105–116.

Arthur, R., 2001, Leibniz on infinite number, infinite wholes, and the whole world: A reply to Gregory Brown, *The Leibniz Review*, 11, 103–116.

Arthur, R., 2015, Leibniz's actual infinite in relation to his analysis of matter, in N. Goethe *et al.* 2015, 137–156.

Baltzer, R., 1870, *Die Elemente der Mathematik*, Leipzig: Hirzel. Third edition.

Beaney, M., 1997, *The Frege Reader*, Oxford: Blackwell.

Bell, J. L., 1998, *A Primer of Infinitesimal Analysis*, Cambridge: Cambridge University Press.

Bellavitis, G., 1835, Saggio di applicazioni di un nuovo metodo di Geometria analitica (Calcolo delle equipollenze), *Ann. Lomb. Veneto*, 5, 244–259.

Belna, J.-P., 2002, Frege et la géométrie projective: La Dissertation inaugurale de 1873, *Revue d'histoire des sciences*, 55, 379–410.

Benci, V., 1995, I numeri e gli insiemi etichettati, *Conferenze del seminario di matematica dell'Universita' di Bari*, vol. 261, Bari: Laterza, 29 pages.

Benci, V. and Di Nasso, M., 2003, Numerosities of labeled sets: A new way of counting, *Advances in Mathematics*, 173, 50–67.

Benci, V., Di Nasso, M., and Forti, M., 2006, An Aristotelean notion of size, *Annals of Pure and Applied Logic*, 143, 43–53.

Benci, V., Di Nasso, M., and Forti, M., 2007, An Euclidean measure of size for mathematical universes, *Logique et Analyse*, 50, 43–62.

Benci, V., Horsten, L., and Wenmackers, S., 2013, Non-Archimedean Probability, *Milan Journal of Mathematics*, 81, 121–151.

Bettazzi, 1887, Sul concetto di numero, *Periodico di matematica per l'insegnamento secondario*, 2, 97–113, 129–143.

Bianchi, L., 1984, *L'Errore di Aristotele. La Polemica Contro l'Eternità del Mondo nel XIII Secolo*, Firenze: La Nuova Italia.

Biard, J. and Celeyrette, J., 2005, *De la Théologie aux Mathématiques. L'Infini au XIVeme Siècle*, Paris: Les Belles Lettres.

Bindoni, A., 1912, Sulle definizioni per astrazione e mediante classi, 11, *Il Bollettino di Matematica*, 153–156.

Black, R., 2000, Nothing matters too much, or Wright is wrong, *Analysis*, 60, 229–237.

Blanchette, P., 2012, *Frege's Conception of Logic*, Oxford: Oxford University Press.

Bolzano, B., 1837, *Wissenschaftslehre*, Sulzbach: Seidel. Partial English translation in Bolzano 1973. Full English translation in Bolzano 2014.

Bolzano, B., 1851, *Paradoxien des Unendlichen*, Leipzig: Reclam. See also Bolzano 1975a. Translated as *Paradoxes of the Infinite*, ed. by Donald A. Steele, London: Routledge and Kegan Paul, and New Haven: Yale University Press, 1950. A more recent translation, which I use, is in Russ 2005.

Bolzano, B., 1973, *Theory of Science*, Dordrecht: Reidel.

Bolzano, B., 1975a, *Paradoxien des Unendlichen*, Hamburg: Felix Meiner Verlag.

Bolzano, B., 1975b, *Einleitung zur Grössenlehre. Erste Begriffe der allgemeinen Grössenlehre, Gesamtausgabe, II A 7*, ed. by Jan Berg, Stuttgart-Bad Cannstatt: Friedrich Fromann Verlag.

Bolzano, B., 1978, *Vermischte philosophische und physikalische Schriften 1832–1848, Gesamtausgabe, II A 12:2*, ed. by Jan Berg, Stuttgart-Bad Cannstatt: Friedrich Fromann Verlag.

Bolzano, B., 2014, *The Theory of Science*, translated by R. George and P. Rusnock, 4 vols., Oxford: Oxford University Press.

Boniface, J., 2002, *Les constructions des nombres réels dans le mouvement d'arithmétisation de l'analyse*, Paris: Ellipses.

Boniface, J., 2007, The concept of number from Gauss to Kronecker, in C. Goldstein *et al.* 2007, 315–342.

Boniface, J. and Schappacher, N., 2001, 'Sur le concept de nombre en mathématique' Cours inédit de Leopold Kronecker à Berlin (1891), *Revue d'histoire des mathématiques*, 7, 207–275.

Boolos, G., 1987, The consistency of Frege's *Foundations*, in J. Thomson, ed., *On Being and Saying: Essays in Honor of Richard Cartwright*, Cambridge MA: MIT Press, 3–20; reprinted in Boolos 1998, 183–201.

Boolos, G., 1990, The standard equality of numbers, in G. Boolos, ed., *Meaning and Method: Essays in honor of Hilary Putnam*, Cambridge: Cambridge University Press, 261–277. Reprinted in Boolos 1998, 202–219.

Boolos, G., 1996, On the proof of Frege's theorem, in A. Morton and S. P. Stich, eds., *Paul Benacerraf and his Critics*, Cambridge MA: Blackwell. Reprinted in Boolos 1998, 275–290.

Boolos, G., 1997, Is Hume's principle analytic?, in Heck 1997b, 245–261; reprinted in Boolos 1998, 301–314.

Boolos, G., 1998, *Logic, Logic, and Logic*, Cambridge MA: Harvard University Press.

Borga, M., Freguglia, P., and Palladino, D., eds., 1985, *I Contributi Fondazionali della Scuola di Peano* Milano: Franco Angeli.

Bos, H., 1974, Differentials, higher-order differentials and the derivative in the Leibnizian calculus, *Archive for History of Exact Sciences*, 14, 1–90.

Bradwardine, T., 1979, *Geometria Speculativa*, Wiesbaden: F. Steiner Verlag.

Brandom, R. B., 1986, Frege's technical concepts: Some recent developments, in L. Haaparanta and J. Hintikka, eds., *Frege Synthesized*, Dordrecht: Reidel, 253–295.

Breger, H., 2008, Natural numbers and infinite cardinal number, in H. Hecht *et al.*, eds., *Kosmos und Zahl*, Stuttgart: Steiner, 309–318.

Brown, G., 2000, Leibniz on wholes, unities, and infinite number, *The Leibniz Review*, 10, 21–51.

Bunn, R., 1977, Quantitative relations between infinite sets, *Annals of Science*, 34, 177–191.

Burali-Forti, C., 1894*a*, Sulle classi ordinate e i numeri transfiniti, *Rendiconti del Circolo Matematico di Palermo*, VIII, 169–179.

Burali-Forti, C., 1894*b*, *Logica Matematica*, Milano: Hoepli.

Burali-Forti, C., 1896*a*, Le classi finite, *Atti dell'Accademia Reale delle Scienze di Torino*, 32, 34–51.

Burali-Forti, C., 1896*b*, Sopra un teorema del sig. G. Cantor, *Atti dell'Accademia Reale delle Scienze di Torino*, 31, 229–237.

Burali-Forti, C., 1899*a*, Sur l'égalité et sur l'introduction des éléments dérivés dans la science, *L'enseignement mathématique*, 1, 246–261.

Burali-Forti, C., 1899*b*, Les propriétés formales des opérations algébriques, *Rivista di Matematica*, 6, 141–177.

Burali-Forti, C., 1901, Sur les différentes methods logiques pour la definition du nombre reel, *Congrès International de Philosophie*, Paris: Colin, 289–307.

Burali-Forti, 1909, Sulle definizioni mediante "coppie", *Il Bollettino di Matematica*, anno VIII, 237–242.

Burali-Forti, C., 1912, Gli enti astratti come enti relativi ad un campo di nozioni, *Rendiconti dell'Accademia dei Lincei*, 21, part II, 677–682.

Burali-Forti, C., 1919, *Logica Matematica*, seconda edizione, Milano: Hoepli.

Burali-Forti, C., 2013, *Logica Matematica*, edited with an introduction by G. Lolli, Pisa: Edizioni della Normale. (The book contains the 1894 edition as well as the 1919 edition.)

Burali-Forti, C. and Marcolongo, R., 1909, *Elementi di Calcolo Vettoriale, con numerose applicazioni alla Geometria, alla Meccanica e alla Fisica-Matematica*, Bologna: Zanichelli.

Burali-Forti, C. and Marcolongo, R., 1910, *Éléments de calcul vectoriel avec de nombreuses application à la géométrie, à la mécanique et à la physique mathématique*, Paris: Hermann. French translation of Burali-Forti and Marcolongo 1909.

Burbage, F. and Chouchan, N., 1993, *Leibniz et l'infini*, Paris: PUF.

Buzaglo, M., 2002, *The Logic of Concept Expansion*, Cambridge: Cambridge University Press.

Cantor, G., 1872, Über die Ausdehnung eines Satzes aus der Theorie der trigonometrischen Reihen, in Cantor 1932, 92–102.

Cantor, G., 1878, Ein Beitrag zur Mannigfaltigkeitslehre, in Cantor 1932, 119–133.

Cantor, G., 1879, Über unendliche lineare Punktmannigfaltigkeiten, in Cantor 1932, 139–246.

Cantor, G., 1883, *Grundlagen einer allgemeinen Mannigfaltigkeitslehre*, Leipzig: Teubner.

Cantor, G., 1887/1888, Mitteilungen zur Lehre vom Transfiniten, *Zeitschrift fur Philosophie und philosophische Kritik*, 91, 81–125 (1887); 92, 240–265 (1888).

Cantor, G., 1895/1897, *Beiträge zur Begründung der transfiniten Mengenlehre*, in Cantor 1932, 282–356.

Cantor, G., 1932 (1962), *Gesammelte Abhandlungen*, Hildesheim: Georg Olms.

Cantor, M., 1901, Nachruf an Oskar Schlömilch, *Bibliotheca Mathematica*, 3. Folge, 2, 260–281.

Carnap, R., 1929, *Abriss der Logistik*, Vienna: Julius Springer.

Cassina, U., 1961, *Critica dei Principi della Matematica e Questioni di Logica*, Roma: Edizioni Cremonese.

Catania, S., 1911, Sulle definizioni per astrazione, *Il Bollettino di Matematica*, 10, 153–156.

Cauchy, A., 1847, *Exercises d'Analyse*, Paris: Bachelier.

Christopolou, D., 2014, Weyl on Fregean implicit definitions: Between phenomenology and symbolic construction, *Journal for General Philosophy of Science*, 45, 35–47.

Cipolla, M., 1914, *Analisi Algebrica e Introduzione al Calcolo Infinitesimale*, Palermo: Capozzi.

Clark, P., 2004, Frege, neo-logicism and applied mathematics, in F. Stadler, ed., *Induction and Deduction in the Sciences*, Dordrecht: Kluwer, 2004. Reprinted in Cook 2007, 45–60.

Cook, R., 2002, The state of the economy: Neo-logicism and inflation, *Philosophia Mathematica* (3), vol 10, 43–66.

Cook, R., ed., 2007, *The Arché Papers on the Mathematics of Abstraction*, Dordrecht: Springer.

Cook, R., 2012, Conservativeness, stability, and abstraction, *British Journal for the Philosophy of Science*, 63, 673–696.

Cook, R. and Ebert, P., 2005, Abstraction and identity, *Dialectica*, 59, 121–139.

Couturat, L., 1896, *De l'Infini Mathématique*, Paris: Alcan.

Couturat, L., 1905, *Les Principes des Mathématiques*, Paris: Alcan.

Cross, R., 1998, Infinity, continuity, and composition: The contribution of Gregory of Rimini, *Medieval Philosophy and Theology*, 7, 89–110.

Cutland, N. J., Di Nasso, M., and Ross, D. A., eds., 2006, *Nonstandard Methods and Applications in Mathematics*, Wellesley, MA: AK Peters.

Dales, R. C., 1984, Henry of Harclay on the infinite, *Journal of the History of Ideas*, 45, 295–301.

Darrigol, O., 2003, Number and Measure: Hermann von Helmholtz at the croassroads of mathematics, physics, and psychology, *Studies in History and Philosophy of Science*, 34, 515–573.

Dauben, J. W., 1990, *George Cantor. His Mathematics and Philosophy of the Infinite*, Princeton: Princeton University Press.

Dauben, J. W., 1995, *Abraham Robinson. The creation of nonstandard analysis. A personal and mathematical Odyssey*, Princeton: Princeton University Press.

De Amicis, E., 1892, Dipendenza fra alcune proprietà notevoli delle relazioni fra enti di un medesimo sistema, *Rivista di Matematica*, 1, 113–127.

Dedekind, R., 1872, *Stetigkeit und irrationale Zahlen*, Braunschweig: Vieweg. English transation in Ewald 1996, 756–79.

Dedekind, R., 1888, *Was sind und was sollen die Zahlen*, Braunschweig: Vieweg. English transation in Ewald 1996, 787–833.

Deiser, O., 2010, On the development of the notion of a cardinal number, *History and Philosophy of Logic*, 31, 123–143.

Demopoulos, W., 2003, On the philosophical interest of Frege Arithmetic, *Philosophical Books*, 44, 220–228. Reprinted in Cook 2007, 105–115.

Desargues, G., 1636, *Brouillon Projet*, edited in original language with facing English translation in J. V. Field and J. J. Gray 1987.

Dewender, T., 2002, *Das Problem des Unendlichen im ausgehenden 14. Jahrhundert. Eine Studie mit Textedition zum Physikkommentar des Lorenz von Lindores*, Amsterdam: B.R. Grüner Publishing Co.

Di Nasso, M. and Forti, M., 2010, Numerosities of point sets over the real line, *Transactions of the American Mathematical Society*, vol. 362, 5355–71.

Dirichlet, P. G., 1863, *Vorlesungen über Zahlentheorie*, edited by R. Dedekind, Braunschweig: Vieweg. (1879, later edition). English translation in Dirichlet 1999.

Dirichlet, P. G., 1999, *Lectures on Number Theory. Supplement by R. Dedekind*, Providence: American Mathematical Society.

Dodgson, C. L., 1885, *Euclid and his Modern Rivals*, second edition (first edition 1879), London: MacMillan.

Dubislav, W., 1926, *Die Definition*, second edition, Leipzig: Meiner.

Dubislav, W., 1931, *Die Definition*, third edition, Leipzig: Meiner.

Duggan, J., 1999, A general extension theorem for binary relations, *Journal of Economic Theory*, 86, 1–16.

Duhem, P., 1955, Léonard de Vinci et les deux infinis, in *Études sur Léonard de Vinci*, Seconde Serie, Paris: De Nobele, 3–53; 368–407.

Dühring, E., 1877, *Kritische Geschichte der allgemeinen Principien der Mechanik*, Leipzig: Fues's Verlag; first edition 1869.

Dummett, M., 1998, Neo-Fregeans; in bad company?, in *The Philosophy of Mathematics Today*, M. Schirn, ed., New York: Clarendon Press, 1988, 369–387.

Dummett, M., 1991, *Frege. Philosophy of Mathematics*, Cambridge MA: Harvard University Press.

Dushnik, B. and Miller, E. W., 1941, Partially ordered sets, *American Journal of Mathematics*, 63, 600–610.

Ebert, P. and Rossberg, M., 2009, Cantor on Frege's *Foundations of Arithmetic*: Cantor's 1885 review of Frege's *Die Grundlagen der Arithmetik*, *History and Philosophy of Logic*, 30, 341–348.

Ehrlich, P., ed., 1994, *Real Numbers, Generalizations of the Reals, and Theories of Continua*, Dordrecht: Kluwer Academic Publishers.

Ehrlich, P., 2006, The rise of non-Archimedean mathematics and the roots of a misconception I: The emergence of non-Archimedean systems of magnitudes, *Archive for History of Exact Sciences*, 60, 2006, 1–121.

Ehrlich, P., 2012, The absolute arithmetical continuum and the unification of all numbers great and small, *The Bulletin of Symbolic Logic*, 18, 1–45.

Enriques, F., 1922*a*, *Per la Storia della Logica*, Bologna: Zanichelli. Translated into English as *The Historic Development of Logic*, New York: H. Holt and Co., 1929.

Enriques, F., 1922*b*, Il positivismo e la critica degli assiomi dell'uguaglianza, *Periodico di Matematiche*, s. IV, vol. II, 185–187.

Etcheverría, J., 1979, L'analyse géométrique de Grassmann et ses rapports avec la Caractéristique Géométrique de Leibniz, *Studia Leibnitiana*, 11, 223–273.

Euclid, 1956, *The Thirteen Books of Euclid's Elements*, edited by T. L. Heath, New York: Dover. (Reprint of the second edition published by Cambridge University Press, 1925).

Euler, L., 1748, *Introductio in Analysin Infinitorum*, vol. II, Lausanne: Bousquet Socios. Translated into English in Euler 1990.

Euler, L., 1990, *Introduction to the Analysis of the Infinite*, vol. II, Berlin: Springer.

Ewald, W., 1996, *From Kant to Hilbert*, 2 vols., Oxford: Oxford University Press.

Felgner, U., 2002, Der Begriff der Kardinalzahl, in Hausdorff 2002, 634–644.

Ferreirós, J., 2007_2, (1999_1), *Labyrinth of Thought: A history of set theory and its role in modern mathematics*, Basel: Birkhäuser. Second edition.

Field, J. V. and Gray, J. J., 1987, *The Geometrical Work of Girard Desargues*, Berlin: Springer Verlag.

Fine, B. and Rosenberger, G., 2007, *Number Theory. An Introduction Through the Distribution of Primes*, Boston: Birkhäuser.

Fine, K., 1998, Cantorian abstraction: A reconstruction and defense, *Journal of Philosophy*, 95, 599–634.

Fine, K., 2002, *The Limits of Abstraction*, Oxford: Oxford University Press.

Fischer, E. G. F., 1820_1, (1833_2), *Lehrbuch der Ebenen Geometrie für Schulen*, Nauck: Berlin und Leipzig.

Flitner, A. and Wittig, J., eds., 2000, *Optik–Technik–Soziale Kultur. Siegfried Czapski, Weggefährte und Nachfolger Ernst Abbes. Briefe, Schriften, Dokumente*, Rudolstadt: Hain-Verlag.

Flood, J., Ginther, J., and Goering, J., eds., 2013, *Robert Grosseteste and His Intellectual Milieu*, Toronto: Pontifical Institute of Mediaeval Studies.

Fraenkel, A., 1927, *Zehn Vorlesungen über die Grundlegung der Mengenlehre*, Leipzig und Berlin: Teubner.

Fraenkel, A., 1928, *Einleitung in die Mengenlehre*, Berlin: Springer. Third edition.

Frege, G., 1873, *Über eine geometrische Darstellung der imaginären Gebilde in der Ebene*, Inaugural-Dissertation der Philosophischen Facultät zu Göttingen zur Erlangung der Doctorwürde vorgelegt, Jena: A. Neuenhan. English translation in Frege 1984, 1–55.

Frege, G., 1877, Rezension von : A. v. Gall und E. Winter, Die Analytische Geometrie des Punktes und der Geraden und ihre Anwendung auf Aufgaben, *Jenaer Literaturzeitung*, 4, 133–134. English translation in Frege 1984, 95–97.

Frege, G., 1884, *Die Grundlagen der Arithmetik. Eine logisch mathematische Untersuchung über den Begriff der Zahl*, Breslau: Koebner; see also Frege 1986. Translated into English by J. Austin in *The Foundations of Arithmetic*, 2nd edition, New York: Harper, 1960.

Frege, G., 1893, *Grundgesetze der Arithmetik I*, Jena: H. Pohle. (Reprint by G. Olms, Hildesheim, 1962).

Frege, G., 1903, *Grundgesetze der Arithmetik II*, Jena: H. Pohle. (Reprint by G. Olms, Hildesheim, 1962).

Frege, G., 1976, *Wissenschaftlicher Briefwechsel*, edited by G. Gabriel, H. Hermes, F. Kambartel, C. Thiel, and A. Veraart, Hamburg: Felix Meiner Verlag. English translation in Frege 1980.

Frege, G., 1979, *Posthumous Writings*, edited by H. Hermes, F. Kambartel, F. Kaulbach, Chicago: University of Chicago Press.

Frege, G., 1980, *Philosophical and Mathematical Correspondence*, edited by B. McGuinness, Oxford: Blackwell.

Frege, G., 1983, *Nachgelassene Schriften*, edited by G. Gabriel, H. Hermes, F. Kambartel, F. Kaulbach, W. Rödding, second revised edition, Hamburg: Felix Meiner Verlag.

Frege, G., 1984, *Collected Papers on Mathematics, Logic and Philosophy*, ed. by B. McGuinness, Oxford: Blackwell.

Frege, G., 1986, *Die Grundlagen der Arithmetik. Eine logisch mathematische Untersuchung über den Begriff der Zahl, Centenarausgabe*, edited by C. Thiel, Hamburg: Felix Meiner Verlag.

Frege, G., 2013, *Basic Laws of Arithmetic*, translated and edited by P. A. Ebert, M. Rossberg, and C. Wright, Oxford: Oxford University Press.

Freguglia, P., 1982, Definizioni per astrazione e numeri cardinali, *Cultura e Scuola*, vol. 81, 199–208.

Freguglia, P., 1992, *Dalle equipollenze ai sistemi lineari. Il contributo italiano al calcolo geometrico*, Urbino: QuattroVenti.

Freguglia, P., 2006, *Geometria e Numeri. Storia, teoria elementare e applicazione del calcolo geometrico*. Torino: Bollati Boringhieri.

Galileo, 1939, *Dialogues Concerning Two New Sciences*, Evanston and Chicago: Northwestern University; reprinted by Dover in 1954.

Galileo, 1958, *Discorsi e Dimostrazioni Intorno a Due Nuove Scienze*, a cura di Adriano Carugo e Ludovico Geymonat, Torino: Boringhieri.

Gardies, J.-L., 1984, *Pascal entre Eudoxe et Cantor*, Paris: Vrin.

Gauss, C. F., 1801, *Disquisitiones Arithmeticae*, Lipsia: Fleische. English translation in C. F. Gauss, *Disquisitiones Arithmeticae*, New York: Springer, 1986.

Gauss, 1816, Review of Schwab 1814, in Gauss 1873, 364–368.

Gauss, C. F., 1873, *Werke*, vol. 4, Göttingen: Königliche Gesellschaft der Wissenschaften. Reprinted by Cambridge University Press, 2011.

Gauss, C. F., 1874, *Werke*, vol. 6, Göttingen: Königliche Gesellschaft der Wissenschaften. Reprinted by Cambridge University Press, 2011.

Gericke, H., 1977, Wie vergleicht man unendliche Mengen?, *Sudhoffs Archiv*, 61, 54–65.

Gilbert, T. and Rouche, N., 1996, Y-at-il vraiment autant de nombres pairs que des naturels?, in A. Pétry, ed., *Méthodes et Analyse Non Standard*, Cahiers du Centre de Logique, Vol. 9, Bruylant-Academia, 99–139.

Giusti, E., 1993, *Euclides reformatus. La teoria delle proporzioni nella scuola galileiana*, Torino: Bollati Boringhieri.

Gödel, K., 1990, *Collected Works*, edited by S. Feferman, J. W. Dawson, S. C. Kleene, G. H. Moore, R. Solovay, and J. van Heijenoort, Vol. II. New York: Oxford University Press.

Goethe, N., Beeley, P., and Rabouin, D., eds., 2015, G. W. *Leibniz, Interrelations between Mathematics and Philosophy*, Berlin: Springer.

Goldblatt, R., 1998, *Lectures on the Hyperreals. An Introduction to Nonstandard Analysis*, New York: Springer.

Goldman, J. R., 2002, *The Queen of Mathematics: A Historically Motivated Guide to Number Theory*, Wellesley, MA: AK Peters.

Goldstein, C., Schappacher, N., and Schwermer, J., eds., 2007, *The Shaping of Arithmetic after C. F. Gauss's Disquisitiones Arithmeticae*, Berlin-Heidelberg-New York: Springer.

Grassmann, H., 1844, *Die Lineale Ausdehnungslehre ein neuer Zweig der Mathematik, dargestellt und durch Anwendungen auf die übrigen Zweige der Mathematik, wie auch auf die Statik, Mechanik, die Lehre vom Magnetismus und die Krystallonomie erläutert*, Leipzig: Otto Wiegand. English translation in Grassmann 1995.

Grassmann, H., 1847, *Geometrische Analyse geknüpft an die von Leibniz erfundene geometrische Charakteristik. Gekrönte Preisschrift*, Leipzig: Wiedmann. English translation in Grassmann 1995.

Grassmann, H., 1861, *Lehrbuch der Arithmetik für höhere Lehranstalten*, Berlin: Enslin.

Grassmann, H., 1995, *A New Branch of Mathematics. The Ausdehnungslehre of 1844 and Other Works*, edited and translated by L. C. Kannenberg, Chicago: Open Court.

Grattan-Guinness, I., 2000, *The Search for Mathematical Roots 1870–1940. Logics, set theories, and the foundations of mathematics from Cantor through Russell and Gödel*, Princeton: Princeton University Press.

Grosseteste, R., 2011, *La luce*, with introduction, Latin text, translation into Italian and commentary by C. Panti, Pisa: Plus-Pisa University Press.

Grunert, J., 1857, Theorie der Aequivalenzen, *Archiv der Mathematik und Physik*, 443–477.

Grunert, J., 1870 [1851$_1$], *Lehrbuch der ebenen Geometrie*, Brandenburg a/H.: Wiesike.

Gugler, B., 1850, Ueber die Begründung der Elementar-Geometrie, *Zeitschrift für das Gesammtschulwesen*, 6, 259–312.

Hale, R., 1997 (2001), *Grundlagen* §64, *Proceedings of the Aristotelian Society*, 97, 243–261. Reprinted in Hale and Wright, 2001a, 91–116.

Hale, R. and Wright, C. 2001a, *The Reason's Proper Study*, Oxford: Clarendon Press.

Hale, R. and Wright, C. 2001b, "To bury Caesar . . .", in Hale and Wright 2001a, 335–396.

Hale, R. and Wright, C., 2009, Focus restored: comments on John MacFarlane, in Linnebo 2009a, 457–482.

Hallett, M., 1984, *Cantorian Set Theory and Limitation of Size*. Oxford: Clarendon Press. Foreword by Michael Dummett. Reprinted in paperback, with revisions, 1986, 1988.

Hamilton, W. R., 1837, Theory of conjugate functions, or algebraic couples; with a preliminary and elementary essay on algebra as the science of pure time, *Transactions of the Royal Irish Academy*, 17, part 1, 293–422.

Hamilton, W. R., 1853, *Lectures on Quaternions*, Dublin: Hodges and Smith.

Hamilton, W. R., 1866, *Elements of Quaternions*, London: Longmans, Green and Co.

Hankel, H., 1867, *Vorlesungen über die Complexen Zahlen und ihre Functionen*, Leipzig: Leopold Voss.

Harclay, H., 2008, *Ordinary Questions*, edited by M. Henninger, 2 vols., Oxford: Oxford University Press.

Hasse, H., 1926, *Höhere Algebra*, Berlin und Leipzig: de Gruyter.

Haupt, O., 1929, *Einführung in die Algebra*, Leipzig: Akademische Verlagsgesellschaft.

Hausdorff, F., 1914, *Grundzüge der Mengenlehre*, Leipzig: Veit.

Hausdorff, F., 1927, *Mengenlehre*, Berlin und Leipzig: Walter de Gruyter & Co. (Third edition: 1935).

Hausdorff, F., 1957, *Set Theory*, translated by J. R. Aumann *et al.*, New York: Chelsea Publishing Company.

Hausdorff, F., 2002, *Gesammelte Werke*, edited by E. Brieskorn *et al.*, *Grundzüge der Mengenlehre*, Band II, Berlin: Springer.

Hausdorff, F., 2008, *Gesammelte Werke*, edited by E. Brieskorn *et al.*, *Deskriptive Mengenlehre und Topologie*, Band III, Berlin: Springer.

Hausdorff, F., 2013, *Gesammelte Werke*, edited by U. Felgner *et al.*, *Allgemeine Mengenlehre*, Band IA, Berlin: Springer.

Heck, R., 1997*a*, Finitude and Hume's principle, *Journal of Philosophical Logic*, 26, 589–617. Reprinted with a postscript in Heck 2011, 237–266.

Heck, R., 1997*b*, *Logic, Language and Thought*, Oxford: Oxford University Press.

Heck, R., 2000, Cardinality, counting, and equinumerosity, *Notre Dame Journal of Formal Logic*, 41, 187–209. Reprinted in Heck 2011, 156–179.

Heck, R., 2011, *Frege's Theorem*, Oxford: Oxford University Press.

Heine, H., 1872, Die Elemente der Functionenlehre, *Journal für die reine und angewandte Mathematik*, 74, 172–188.

Helmholtz, H., 1887, Zählen und Messen, erkenntnistheoretisch betrachtet, in *Philosophische Aufsätze, Eduard Zeller zu seinem fünfzigjährigen Doctorjubiläum gewidmet*, Leipzig: Fues's Verlag, 17–52. English translation (Counting and Measuring) in Ewald 1996, vol. II, 727–752.

Hermann, K., 2000, *Mathematische Naturphilosophie in die Grundlagendiskussion Jakob Friedrich Fries und die Wissenschaften*, Göttingen: Vandenhoeck & Ruprecht.

Hilbert, D., 2004, *David Hilbert's Lectures on the Foundations of Geometry, 1891–1902*, Michael Hallett and Ulrich Majer, editors. Berlin: Springer-Verlag.

Hindenburg, C. F., 1781, Ueber die Schwürigkeit bey der Lehre von den Parallellinien. Neues System der Parallellinien, *Leipziger Magazin für Naturkunde, Mathematik und Oekonomie*. Zweiter Stük [sic!], 145–168.

Husserl, E., 1891, *Philosophie der Arithmetik*, Halle: Pfeffer.

Israel, J., 2002, *Radical Enlightenment: Philosophy and the Making of Modernity*, Oxford: Oxford University Press.

Jacobi, A., 1824, *De undecimo Euclidis Axiomate*, Jena: Kroecker.

Kamke, E., 1928, *Mengenlehre*, Berlin und Leipzig: W. de Gruyter.

Kanamori, A., 1997, The mathematical import of Zermelo's well-ordering theorem, *The Bulletin of Symbolic Logic*, 3, 281–311.

Karsten, W. J. G., 1778, *Versuch einer völlig berichtigten Theorie von den Parallellinien*, Halle.

Katz, F. M., 1981, *Sets and their Sizes*, Ph.D. Dissertation, MIT. Now newly typeset (2001) and downloadable at http://citeseerx.ist.psu.edu/viewdoc/summary?doi=10.1.1.28.7026

Killing, W., 1893, *Einführung in die Grundlagen der Geometrie*, vol. I, Paderborn: Schöningh.

Kirschner, S., 1997, *Nicolaus Oresmes Kommentar zur Physik des Aristoteles. Kommentar mit Edition der Quaestiones zu Buch 3 und 4 der aristotelischen Physik sowie von vier Quaestiones zu Buch 5*, Wiesbaden: Franz Steiner Verlag, (Sudhoff Archiv, Beihefte 39).

Kitcher, P., 1984, *The Nature of Mathematical Knowledge*, Oxford: Oxford University Press.

Knobloch, E., 1999, Galileo and Leibniz: Different approaches to infinity, *Archive for History of Exact Sciences*, 54, 87–99.

Kossak, E., 1872, *Die Elemente der Arithmetik*, Berlin: Nauck.

Kreiser, L., 1984, 'G. Frege "Die Grundlagen der Arithmetik" Werk und Geschichte', in G. Wechsung, ed., *Frege Conference 1984* (Proceedings of the International Conference Held at Schwerin, GDR, September 10–14, 1984), Berlin: Akademie-Verlag, 13–27.

Kreiser, L., 2001, *Gottlob Frege. Leben-Werk-Zeit*, Hamburg: Felix Meiner Verlag.

Kronecker, L., 1882, *Grundzüge einer arithmetischen Theorie der algebraischen Grössen*, Berlin: Reimer. Reproduced in Kronecker 1895–1930, vol. II, 239–387.

Kronecker, L., 1887, Über den Zahlbegriff, *Crelle Journal für die reine und angewandte Mathematik*, 101, 337–355; and in *Philosophische Aufsätze, Eduard Zeller zu seinem fünfzigjährigen Doctor-Jubiläum gewidmet*, Leipzig 1887, 261–274. Reproduced in Kronecker 1895–1930, vol. III, 251–274.

Kronecker, L., 1889–1890, Zur Theorie der elliptischen Functionen, in Kronecker 1895–1930, vol. V, 58–60.

Kronecker, L., 1891, *Über den Begriff der Zahl in der Mathematik*, see Boniface and Schappacher 2001.

Kronecker, L., 1895–1930, *Werke*, edited by K. Hensel, 5 vols., Leipzig: Teubner. Reprinted in 1968 by Chelsea, New York.

Leibniz, G. W., 1875–1890, *Die philosophischen Schriften*, edited by C. I. Gerhardt, Berlin: Weidmann. Reprint by G. Olms.

Leibniz, G. W., 1955 (and other editions), *The Leibniz-Clarke Correspondence*, edited by H. G. Alexander, Manchester: Manchester University Press. See also Leibniz 2000.

Leibniz, G. W., 2000, *G. W. Leibniz and Samuel Clarke: Correspondence*, edited, with introduction, by Roger Ariew, Indianapolis/Cambridge: Hackett Publishing Co. Inc.

Leibniz, G. W., 2001, *The Labyrinth of the Continuum*, New Haven: Yale University Press.

Levey, S., 2015, Comparability of infinities and infinite multitudes in Galileo and Leibniz, in Goethe *et al.* 2015, 157–187.

Lévy, T., 1987, *Figures de l'infini*, Paris: Seuil.

Lewis, N., 2012, Robert Grosseteste and Richard Rufus of Cornwall on unequal infinites, in J. P. Cunningham, ed., *Robert Grosseteste His Thought and Its Impact*, Toronto: Pontifical Institute of Mediaeval Studies, 2012, 227–256.

Lewis, N., 2013*a*, Robert Grosseteste, *Stanford Encyclopedia of Philosophy*; http://plato.stanford.edu/entries/grosseteste.

Lewis, N., 2013*b*, Robert Grosseteste's On Light: An English Translation, in J. Flood *et al.* 2013, 239–247. (A translation of *De luce* based on the critical edition in Panti 2013.)

Lipschitz, R., 1986, *Briefwechsel*, edited by W. Scharlau, Braunschweig: Vieweg.

Lieber, H. W. and von Lühmann, F., 1876, *Leitfaden der Elementarmathematik. Erster Teil. Planimetrie*, Berlin: Simion. (First edition).

Lieber, H. W. and von Lühmann, F., 1902, *Leitfaden der Elementarmathematik. Erster Teil. Planimetrie*, Berlin: Simion. Ninth edition revised by Carl Müsebeck.

Linnebo, Ø., ed., 2009*a*, The Bad Company Problem, *Synthese*, Special Issue, Vol. 170, No. 3, October 2009.

Linnebo, Ø., 2009*b*, Introduction, in Linnebo 2009*a*, 321–329.

Linnebo, Ø., and Pettigrew, R., 2014, Two types of abstraction for structuralism, *The Philosophical Quarterly*, 64, 267–283.

Lolli, G., 2013, Cesare Burali-Forti (1861–1931) e la logica matematica, in Burali-Forti 2013, vii–lxiii.

Lorey, A., 1868, *Lehrbuch der ebenen Geometrie nach genetisch-heuristischer Weise*, Vera und Leipzig: Kanitz Verlag.

Lüroth, J., 1881, *Grundriss der Mechanik*, München: Theodor Ackermann.

MacBride, F., 2000, On Finite Hume, *Philosophia Mathematica* (3), 8, 150–159.

MacBride, F., 2002, Could nothing matter?, *Analysis*, 62, 125–135. Reprinted in Cook 2007, 95–104.

MacBride, F., 2003, Speaking with shadows: A study of neo-logicism, *British Journal for Philosophy of Science*, 54, 103–163.

Maccaferri, E., 1913, Le definizioni per astrazione e la classe di Russell, *Rendiconti del Circolo Matematico di Palermo*, 165–171.

Mago, V., 1913, Teoria degli ordini, *Memorie della Reale Accademia delle scienze di Torino* (2), 64, 1–25.

Maier, A., 1949, *Die Vorläufer Galileis im 14. Jahrhundert; Studien zur Naturphilosophie der Spätskolastik*. Rome: Edizioni di Storia e Letteratura.

Maignan, E., 1673_2, *Cursus philosophicus*, Lyon: Johannis Grégoire. (1st ed., 4 vols., Toulouse, 1652; 2nd ed. with changes and additions, Lyon, 1673).

Mancosu, P., 1996, *Philosophy of Mathematics and Mathematical Practice in the Seventeenth Century*, Oxford: Oxford University Press.

Mancosu, P., 2008*a*, Mathematical explanation: Why it matters, in P. Mancosu, 2008*b*, 134–149.

Mancosu, P., ed., 2008*b*, *The Philosophy of Mathematical Practice*, Oxford: Oxford University Press.

Mancosu, P., 2009, Measuring the size of infinite collections of natural numbers: Was Cantor's theory of infinite number inevitable?, *The Review of Symbolic Logic*, 2, 612–646.

Mancosu, P., 2015*a*, *Grundlagen*, Section 64: Frege's discussion of definitions by abstraction in historical context, *History and Philosophy of Logic*, 36, 2015, 62–89.

Mancosu, P., 2015*b*, In Good Company? On Hume's Principle and the assignment of numbers to infinite concepts, *The Review of Symbolic Logic*, 8, issue 2, 2015, 370–410. (A version in French appeared in P. Mancosu, 2015*c*, pp. 73–122).

Mancosu, P. 2015*c*, *Infini, Logique, Géométrie*, Paris: Vrin.

Mancosu, P., forthcoming, Definitions by abstraction in the Peano school, in C. de Florio and A. Giordani, eds., *From Arithmetic to Metaphysics. A Path through Philosophical Logic. Studies in honor of Sergio Galvan*, Berlin: de Gruyter.

Mancosu, P. and E. Vailati, E., 1991, Torricelli's infinitely long solid and its philosophical reception in the XVIIth century, *ISIS*, 82, 50–70.

Meschowski, H., 1973, *Hundert Jahre Mengenlehre*, Munich: Deutscher Taschenbuch Verlag.

Meulders, M., 2010, *Helmholtz. From Enlightenment to Neuroscience*, Cambridge (Mass.): MIT Press.

Monnoyeur, F., ed., 1992, *Infini des Mathématiciens, Infini des Philosophes*, Paris: Belin.

Moore, A. W., 1990 [2001_2], *The Infinite*, London: Routledge.

Murdoch, J., 1981*a*, Henry of Harclay and the infinite, in A. Maierù and A. Paravicini Bagliani, eds., *Studi sul XIV secolo in memoria di Anneliese Maier*, Roma: Edizioni di storia e letteratura, 219–261.

Murdoch, J., 1981*b*, Mathematics and infinity in the later middle ages, in D. O. Dahlstrom, D. T. Ozar, and L. Sweeney, eds., *Infinity, Proceedings of the American Catholic Philosophical Association*, Vol. 55, Washington D. C., 40–58.

Murdoch, J., 1982, Infinity and Continuity, in N. Kretzmann, A. Kenny, and J. Pinborg, eds., *The Cambridge History of Later Medieval Philosophy*, Cambridge: Cambridge University Press, 564–592.

Natucci, A., 1923, *Il Concetto di Numero e le sue Estensioni*, Torino: Bocca.

Netto, E., 1896, Über die arithmetisch-algebraischen Tendenzen Leopold Kronecker's, *Mathematical papers read at the international mathematical Congress, (Chicago, 1893)*, New York: Macmillan, 243–252.

Padoa, A., 1908, Dell'astrazione matematica, in *Questioni Filosofiche*, Modena: Formignini, 91–104.

Panti, C., 2013, Robert Grosseteste's *De luce*: A Critical Edition, in J. Flood *et al.* 2013, 193–238.

Parker, M., 2009, Philosophical method and Galileo's paradox of infinity, in B. van Kerkhove, ed., *New Perspectives on Mathematical Practices*, Hackensack, NJ: World Scientific, 76–113.

Parker, M., 2013, Set size and part-whole principle, *The Review of Symbolic Logic*, 6, 589–612.

Pascal, J., 2005, Anamorphoses et visions miraculeuses du père Maignan (1602–1676), *MEFRIM: Mélanges de l'École française de Rome : Italie et mediterranée*, 117, 1, 45–71.

Pasch, M., 1882, *Vorlesungen über Neuere Geometrie*, Leipzig: Teubner.

Peano, G., 1888, *Calcolo geometrico secondo l'Ausdehnungslehre di H. Grassmann*, Torino: Bocca.

Peano, G., 1891, Sul concetto di numero, *Rivista di Matematica*, 1, 87–102, 256–267.

Peano, G., 1894, *Notations de Logique Mathématique. Introduction au Formulaire de mathématique*, Torino: Bocca.

Peano G., 1895, Dr. Gottlob Frege, *Grundgesetze der Arithmetik, begriffsschriftlich abgeleitet.*, Erster Band, Jena, 1893, pag. XXXII + 254, *Rivista di Matematica*, v. 5, 122–128.

Peano, G., 1899a, *Formulaire de Mathématiques, publié par la Revue de Mathématiques. t. II, n. 3. Logique mathématique. Arithmétique. Limites. Nombres complexes. Vecteurs. Dérivées. Intégrales.* Torino: Bocca.

Peano, G., 1899b, Sui numeri irrazionali, *Rivista di Matematica*, 6, 126–140.

Peano, G., 1900a, Formules de logique mathématique, *Rivista di Matematica*, 7, 1–41

Peano, G., 1901, Les définitions mathématiques, in *Congrès International de Philosophie*, Paris: Colin, 279–288.

Peano, G., 1901b, *Formulaire de Mathématiques*, t. III, Torino: Bocca.

Peano, G., 1901c, Dizionario di matematica. Parte I, Logica Matematica, *Rivista di Matematica*, die 338 [4 Dicembre], 7, 160–172. Also: *Dizionario di matematica. Parte I, Logica Matematica*, Torino: Gerbone, 9 Dicembre 1901, 15 pages.

Peano, G., 1915, Le definizioni per astrazione, *Bollettino della Mathesis*, VII, 106–120.

Peano, G., 1917–1918, Eguale, (Dal supplemento al Dizionario di cognizioni utili, UTET: Torino, 1917), *Il Bollettino di matematica. Giornale scientifico-didattico per l'incremento degli studi matematici nelle scuole medie (A. Conti)*, 15, 195–198.

Petri, B. and Schappacher, N., 2007, On Arithmetization, in C. Goldstein *et al.* 2007, 343–374.

Petruzzellis, N., 1968, L'infinito nel pensiero di S. Tommaso e di G. Duns Scoto, in *De Doctrina Ioannis Duns Scoti, Acta Congr. Scotistici; Studia Scotistica 2*, vol. 2, Rome, 1968, 435–445.

Petsche, H.-J., 2009, *Hermann Grassmann-Biography*, Basel: Birkhäuser.

Pines, S., 1968, Thabit B. Qurra's conception of number and theory of the mathematical infinite, in *Actes du Onzième Congrès International d'Histoire des Sciences* Sect. III: Histoire des Sciences Exactes (Astronomie, Mathématiques, Physique) (Wroclaw, 1963), 160–166.

Proclus, 1992, *A Commentary on the First book of Euclid's Elements*, ed. by G. Morrow, Princeton: Princeton University Press.

Purkert, W., 1987, *Georg Cantor, 1845–1918*, Basel: Birkhäuser.

Rabinovitch, N., 1970, Rabbai Hasdai Crescas (1340–1410) on numerical infinities, *Isis*, 61, 224–230.

Rashed, M., 2009, Thābit ibn Qurra sur l'existence et l'infini : les Réponses aux questions posées par Ibn Usayyid, in R. Rashed, ed., *Thâbit ibn Qurra : Science and Philosophy in Ninth-Century Baghdad*, Berlin and New York: Walter de Gruyter, 2009, 619–674.

Reck, E., 2003, Dedekind's structuralism: an interpretation and defense, *Synthese*, 137, 369–419.

Reich, K., 1996, The emergence of vector calculus in physics: the early decades, in G. Schubring, ed., *Hermann Günther Grassmann (1809–1877): Visionary mathematician, scientist and neohumanist scholar*, Dordrecht: Kluwer, 1996, 197–210.

Reuschle, 1850, Ueber die synthetische Methode in der Mathematik mit Rücksicht auf Schlömilchs "Gründzüge einer wissenschaftlichen Darstellung der Geometrie des Maßes", Eisenach 1849, *Zeitschrift für das Gesammtschulwesen*, 6, 231–258.

Reye, T., 1866, *Die Geometrie der Lage. Vorträge*, vol. I, Hannover: Rümpler. English translation in Reye 1898.

Reye, T., 1898, *Lectures on the Geometry of Position*, London: MacMillan and Company.

Risi, de, V., 2007, *Geometry and Monadology. Leibniz's Analysis Situs and Philosophy of Space*, Basel-Boston-Berlin: Birkäuser.

Rodriguez-Consuegra, F. A., 1991, *The Mathematical Philosophy of Bertrand Russell: origins and development*, Basel: Birkhäuser.

Roero, C. S., 2010, *Peano e la sua scuola*, Torino: Deputazione subalpina di Storia Patria.

Rumfitt, I., 2001, Hume's principle and the number of objects, *Noûs*, 35, 515–541.

Russ, S., 2005, *The Mathematical Works of Bernard Bolzano*, Oxford: Oxford University Press.

Russell, B., 1901, Sur la logique de relations avec des applications à la théorie des series, *Rivista di Matematica*, VII, 115–148. Reprinted in Russell 1993, 613–627. English translation in Russell 1956 and Russell 1993.

Russell, B., 1903, *The Principles of Mathematics*, Cambridge: Cambridge University Press.

Russell, B., 1914, *Our Knowledge of the External World as a Field for Scientific Method in Philosophy*, Chicago and London: Open Court.

Russell, B., 1956, *Logic and Knowledge; Essays 1901–1950*, edited by R. C. Marsh, London: George Allen and Unwin.

Russell, B., 1993, *Towards the "Principles of Mathematics"*, vol. III of the *Collected Papers of Bertrand Russell*, ed. by G. Moore, London and New York: Routledge.

Russell, B. and Whitehead, A. N., 1910–1913, *Principia Mathematica*, 3 volumes, Cambridge: Cambridge University Press.

Sabra, A., 1997, Thabit ibn Qurra on the infinite and other puzzles; edition and translation of his discussions with ibn Usayyid, *Zeitschrift für Geschichte der Arabisch-Islamischen Wissenschaften*, 11, 1–33.

Saguens, J., 1703, *Ioannes Philosophia Maignani scholastica sive in formam concinniorem et auctiorem scholasticam digesta, distributa in tomos quatuor*, Tolosae: Antonium Pech.

Salanskis, J. M., 1999, *Le Constructivisme Non Standard*, Villeneuve d'Ascq: Presses Universitaires du Septentrion.

Salanskis, J. M. and Sinaceur, H., eds., 1992, *Le Labyrinthe du Continu*, Paris: Springer.

Sarnowsky, J., 1989, *Die aristotelisch-scholastische Theorie der Bewegung. Studien zum Kommentar Alberts von Sachsen zur Physik des Aristoteles*, Münster: Aschendorff.

Schäffer, H., 1857, *Lehrbuch der Stereometrie* (Volume 3 of *Lehrbuch der Geometrie* by K. Snell), Leipzig: Brockhaus.

Schlegel, V., 1872, *System der Raumlehre. Nach der Prinzipien der Grassmann'schen Ausdehnungslehre und als Enleitung in dieselbe*, Lepzig: Teubner.

Schlömilch, O., 1845, *Handbuch der algebraischen Analysis*, Jena: Frommann. (Six editions until 1881)

Schlömilch, O., 1849, *Grundzüge einer wissenschaftlichen Darstellung der Geometrie des Maasses*, Eisenach: Baerecke. (Reprinted seven times until 1888).

Schlömilch, O., 1873_2, *Übungsbuch zum Studium der höheren Analysis*, Leipzig: Teubner. First edition 1868.

Schmid, A.-F., ed., 2001, *Bertrand Russell. Correspondance sur la philosophie, la logique et la politique avec Louis Couturat (1897–1913)*, Paris: Editions Kimé.

Scholz, H. and Schweitzer, H., 1935, *Die sogennanten Definitionen durch Abstraktion*, Leipzig: Felix Meiner.

Schotten, H., 1890, *Inhalt und Methode des planimetrischen Unterrichts. Eine vergleichende Planimetrie*, Leipzig: Teubner.

Schröder, E., 1873, *Lehrbuch der Arithmetik und Algebra*, Leipzig: Teubner.

Schuster, P., Berger, U., and Osswald, H., eds., 2001, *Reuniting the Antipodes. Constructive and Nonstandard Views of the Continuum*, Dordrecht: Kluwer.

Schwab, J. C., 1801, *Tentamen novae parallelarum theoriae notione situs fundatae*, Stuttgart: Libreria Erhardiana.

Schwab, J. C., 1814, *Commentatio in primum elementorum Euclidis librum*, Stuttgart: Steinkopf.

Scott, C. A., 1900, On von Staudt's Geometrie der Lage, *The Mathematical Gazette*, 1, 307–314.

Scotus, D., 1639, *In VIII. Libros Physicorum Aristotelis Quaestiones*, Lugdunii: Laurentii Durand. Reprinted by Georg Olms.

Sebestik, J., 1992, La paradoxe de la réflexivité des ensembles infinis: Leibniz, Goldbach, Bolzano, in Monnoyeur 1992, 175–191.

Sebestik, J., 2002, *Logique et Mathématique chez Bernard Bolzano*, Paris: Vrin.

Sesiano, J., 1988, On an algorithm for the approximation of surds from a Provençal treatise, in C. Hay, *Mathematics from Manuscript to Print, 1300–1600*, Oxford: Clarendon Press, 30–56.

Sesiano, J., 1996, Vergleiche zwischen unendlichen Mengen bei Nicolas Oresme, in M. Folkerts, ed., *Mathematische Probleme im Mittelater—der lateinische und arabische Sprachbereich*, Wiesbaden: Harrassowitz Verlag, 361–378.

Shapiro, S., 2000, Frege meets Dedekind: A neologicist treatment of real analysis, *Notre Dame Journal of Formal Logic*, 41, 335–364.

Shapiro, S. and Ebert, P., 2009, The good, the bad, and the ugly, in Linnebo 2009*a*, 415–441.

Shapiro, S. and Weir, A., 2000, 'Neologicist' logic is not epistemically innocent, *Philosophia Mathematica*, 8, 160–189.

Snell, K., 1841, *Lehrbuch der Geometrie*, Leipzig: Brockhaus.

Sohncke, L. A., 1838, Parallelen, in *Allgemeine Encyclopädie der Wissenschaften und Künste*, edited by J. Hersch and J. Gruber, Leipzig: Brockhaus, 368–384.

Somov, I. I., 1878–79, *Theoretische Mechanik*, Leipzig: Teubner.

Sorabji, R., 1983, *Time Creation and the Continuum*, Cornell: Cornell University Press.

Spalt, D., 1990, Die Unendlichkeit bei Bernard Bolzano, in G. König, ed., *Konzepte des mathematischen Unendlichen im 19. Jahrhundert*, Göttingen: Vandenhoeck & Ruprecht, 189–218.

Stolz, O., 1885, *Vorlesungen über allgemeine Arithmetik*, Leipzig: Teubner.

Stolz, O., and Gmeiner, J. A., 1901, *Theoretische Arithmetik*, Leipzig: Teubner.

Szpilrajn, E., 1930, Sur l'extension de l'ordre partiel, *Fundamenta Mathematicae*, 16, 386–389.

Tannery, J., 1894, *Leçons d'arithmétique, théorique et pratique*, Paris: Colin.

Tapp, C., 2005, *Kardinalität und Kardinale. Wissenschaftshistorische Aufarbeitung der Korrespondenz zwischen Georg Cantor und katholischen Theologen seiner Zeit*, Wiesbaden: Franz Steiner Verlag.

Tappenden, J., 2011, A primer on Ernst Abbe for Frege Readers, in Truth and Values, Essays for Hans Herzberger, *Canadian Journal of Philosophy*, Supplementary volume 34, 31–118.

Tappenden, J., forthcoming, *Philosophy and the Emergence of Contemporary Mathematics: Frege in his Mathematical Context*, Oxford: Oxford University Press.

Tennant, N., 2013, Logicism and Neologicism, in E. Zalta, ed., *Stanford Encyclopedia of Philosophy*.

Terski, F., 2006, *L'anamorphose murale de la Trinité des Monts à Rome: ou l'invisible intelligible*, Montpellier: Editions de l'Esperou.

Textor, M., forthcoming, Concept Words, Predicates and 'a great faultline in Frege's philosophy' (version dated March 2012).

Überweg, F., 1874, *System der Logik und Geschichte der logischen Lehre*, Bonn: Adolph Markus. Fourth edition.

Vailati, G., 1892, Dipendenza fra le proprietà delle relazioni, *Rivista di Matematica*, II, 161–164.

Vailati, G., 1894, Recensione a C. Burali-Forti, *Logica Matematica*, Rivista di Matematica, IV, 143–146.

van Atten, M., 2011, A note on Leibniz's argument against infinite wholes, *The British Journal for the History of Philosophy*, 19, 121–129.

Van der Waerden, B., 1930, *Abstrakte Algebra*, Berlin: Springer.

Van Swinden, J. H., 1834, *Elemente der Geometrie*, Jena: Frommann.

Veraart, A., 1976, Geschichte des wissenschaftlichen Nachlasses Gottlob Freges und seiner Edition. Mit einem Katalog des ursprünglichen Bestands der nachgelassenen Schriften Freges, in M. Schirn, ed., *Studien zu Frege I. Logik und Philosophie der Mathematik*, Stuttgart-Bad Cannstatt: Frommann-Holzboog, 49–106.

Vermehren, C. C. H., 1816, *Versuch die Lehre von den parallelen und convergenten Linien aus einfachen Begriffen vollständig herzuleiten und gründlich zu erweisen*, Güstrow: Ebert.

von Neumann, J., 1923, Zur Einführung der transfiniten Zahlen, *Acta Litterarum ac Scientiarum Regiae Universitatis Hungaricae*, 199–208.

Von Staudt, G. K. C., 1847, *Geometrie der Lage*, Nürnberg: Verlag der Dr. Korn'schen Buchhandlung.

Vuillemin, J., 1966, L'élimination des définitions par abstraction chez Frege, *Revue Philosophique de la France et de l'étranger*, 156, 19–40.

Vuillemin, J., 1971, *La Logique et le Monde Sensible. Étude sur les théories contemporaines de l'abstraction*, Paris: Flammarion.

Weber, H., 1896, *Lehrbuch der Algebra*, vol. II, Braunschweig: Vieweg.

Weber, H., 1906, Elementare Mengenlehre, *Jahresbericht der Deutschen Mathematiker-Vereinigung*, XV, 173–184.

Weierstrass, K., 1988, *Einleitung in die Theorie der analytischen Funktionen (Vorlesung Berlin 1878)*, P. Ullrich, ed., Braunschweig: Vieweg.

Weir, A., 2004, Neo-Fregeanism: An embarrassment of riches, *Notre Dame Journal of Formal Logic*, 44, 13–48. Reprinted in Cook 2007, 383–420.

Weyl, H., 1910, Über die Definitionen der mathematischen Grundbegriffe, *Mathematisch-naturwissenschaftliche Blätter*, 7, 93–95 + 109–113; also in Weyl 1968, vol. I, 298–304.

Weyl, H., 1927, *Philosophie der Mathematik und Naturwissenschaft*, München und Berlin: Oldenbourg Verlag.

Weyl, H., 1968, *Gesammelte Abhandlungen*, Berlin: Springer.

Wilson, M., 1992, Frege : The Royal Road from Geometry, *Noûs*, 26, 149–180. Reprinted in W. Demopoulos, *Frege's Philosophy of Mathematics*, Cambridge-London: Harvard University Press, 1995, 108–159.

Wright, C. 1983, *Frege's Conception of Numbers as Objects*, Aberdeen: Aberdeen University Press.

Wright, C., 1997, Is Hume's principle analytic?, in Heck 1997*b*, 201–244. Reprinted in Hale and Wright 2001*a*, 272–306.

Wright, C., forthcoming, Abstraction and epistemic entitlement: On the epistemological status of Hume's principle, in P. Ebert and M. Rossberg, eds., *Abstractionism*, Oxford: Oxford University Press.

Zellini, P., 2005, *A Brief History of Infinity*, London: Penguin Global.

Zetzsche, K. E., 1870, *Leitfaden für den Unterricht in der Ebenen und Räumlichen Geometrie*, Chemnitz: Brunner.

Zetzsche, K. E., 1878, *Katechismus der Ebenen und Räumlichen Geometrie*, Leipzig: Weber.

Name Index